ELEMENTARY ZOOLOGY

Parent Lesson Planner
(PLP)

 Weekly Lesson Schedule

 Student Worksheets

 Quizzes & Test

Answer Key

4th — 6th grade

1 Year
Science

First printing: October 2013

Master Books®, P.O. Box 726, Green Forest, AR 72638

Master Books® is a division of the New Leaf Publishing Group, Inc.

ISBN: 978-0-89051-724-6

Unless otherwise noted, Scripture quotations are from the New King James Version of the Bible.

Printed in the United States of America

Please visit our website for other great titles:
www.masterbooks.net

For information regarding author interviews,
please contact the publicity department at (870) 438-5288

Master
Books®
A Division of New Leaf Publishing Group
www.masterbooks.net

Where Creation Inspires Education

Since 1975, Master Books has been providing educational resources based on a biblical worldview to students of all ages. At the heart of these resources is our firm belief in a literal six-day creation, a young earth, the global Flood as revealed in Genesis 1–11, and other vital evidence to help build a critical foundation of scriptural authority for everyone. By equipping students with biblical truths and their key connection to the world of science and history, it is our hope they will be able to defend their faith in a skeptical, fallen world.

If the foundations are destroyed, what can the righteous do?
Psalm 11:3; NKJV

As the largest publisher of creation science materials in the world, Master Books is honored to partner with our authors and educators, including:

Ken Ham of Answers in Genesis

Dr. John Morris and Dr. Jason Lisle of the Institute for Creation Research

Dr. Donald DeYoung and Michael Oard of the Creation Research Society

Dr. James Stobaugh, John Hudson Tiner, Rick and Marilyn Boyer, Dr. Tom Derosa, and so many more!

Whether a pre-school learner or a scholar seeking an advanced degree, we offer a wonderful selection of award-winning resources for all ages and educational levels.

*But sanctify the Lord God in your hearts, and always be ready
to give a defense to everyone who asks you a reason for the hope
that is in you, with meekness and fear.*
1 Peter 3:15; NKJV

Permission to Copy

Lessons for a 36-week course!

Overview: This *Elementary Zoology PLP* contains materials for use with *The World of Animals, Dinosaur Activity Book, The Complete Aquarium Adventure,* and *The Complete Zoo Adventure.* Materials are organized by each book in the following sections:

📖	Study guide worksheets
Q	Quizzes
T	Optional Final Exam
🔑	Answer Key

Features: Each suggested weekly schedule has three to five easy-to-manage lessons which combine reading, worksheets, and vocabulary-building opportunities. Worksheets, quizzes, and tests are perforated and three-hole punched – materials are easy to tear out, hand out, grade, and store. As always, you are encouraged to adjust the schedule and materials as you need to in order to best work within your educational program.

Workflow: Students will read the pages in their book and then complete each section of the course materials. Tests are given at regular intervals with space to record each grade. Younger students may be given the option of taking open-book tests.

Lesson Scheduling: Space is given for assignment dates. There is flexibility in scheduling. For example, the parent may opt for a M-W schedule rather than a M, W, F schedule. Each week listed has five days but due to vacations the school work week may not be M-F. Please adapt the days to your school schedule. As the student completes each assignment, he/she should put an "X" in the box.

🕐	Approximately 30 to 45 minutes per lesson, three to five days a week	A great selection of books and authors that include:
🔑	Includes answer keys for worksheets, quizzes, and final exam	***The Complete Zoo Adventure*** - Dr. Gary E. Parker is a popular homeschool author and speaker, with multiple degrees, and a co-founder of Creation Adventures Museum. Mary Parker is a phenomenal amateur paleontologist, has participated in fossil digs around the world.
📖	Worksheets for *World of Animals.*	
🔄	Quizzes are included to help reinforce learning and provide assessment opportunities; optional final exam included	***The Complete Aquarium Adventure*** - Bill and Merilee Clifton are founders of Science Partners, offering creation-based science classes for home school students. For over 30 years, Bill taught various science courses for public, Christian, and home schools and was head of SeaWorld Orlando's Education Department.
📄	Designed for grades 4 to 6 in a one-year course	***Dinosaur Activity Book*** - Earl and Bonita Snellenberger are graduates of Herron School of Art of Indiana University. They have written and illustrated a number of unique, fun and educational projects.
⚗️	Supports activity-based learning	***The World of Animals*** – Martin Walters and Jinny Johnson

Contents

Course Description

NOTE: In using the following course materials, you will need to follow the course calendar – especially for the two Complete Adventure books. Like the *Dinosaur Activity Book*, they are included to provide learning activities and additional material that enhance the study of the *World of Animals*. When referencing the activity pages on the PLP course calendar after the devotionals in the Complete Adventure books have been read, it gives a page range – as always, flexibility is given for you to choose as many or as few of these activities you would like your young learner to do. You can make as many copies of these as you may need.

While the calendar doesn't use all the material in the Complete Adventure books, you can use the additional reading material as additional learning or reading opportunities for your child. And you can also choose to use some of these activities on the days when the course calendar doesn't have activities or reading already specified. The material can also be enjoyed by younger or older students as well if you are planning a family trip to a local zoo or aquarium.

The World of Animals

The World of Animals investigates and describes the anatomy, behavior, and habitats of over 1,000 animals. From microscopic worms and insects to reptiles, birds, and mammals, this book will provide children with an insight into the incredible range of life in God's wonderful world. This comprehensive but easy-to-use book boasts a wide range of features to help bring its readers face-to-face with the science and beauty of the living world. This beautiful book includes up-to-date information on endangered species and environments, obscure facts and animal records, and amazing comparisons.

The Complete Adventure Series

Chart your course and weigh anchor on an amazing educational adventure! Learn more about incredible sea creatures and some of the unique animals you can see at the zoo! Whether you are going on a fun family outing, a field trip, or are learning at home, these books are the perfect mix of important details and fun activities to engage the interest of any child! Deepen your understanding of the animal world today as you learn more about God's fascinating creations! Before, during, or after your adventure, there are helpful insights and included materials to make learning easy!

The Wonder of God's World Dinosaur Activity Book

Enjoy hours of educational fun with *The Wonder of God's World Dinosaur Activity Book*! Assemble a *Protoceratops* patterning toy! Create a *Deinonychus* jumping jack! Craft a *Tyrannosaurus* hand puppet! Play the dinosaur migration game! Educational and entertaining, the Activity Book includes mazes, puzzles, word finds, games, and other skill challenges. Create your own dinosaur mini-movies, solve 15 challenging tangram puzzles, and more! Dozens of hands-on and skill-building activities are included for a variety of age levels. Learn the history of dinosaur discoveries and about the lives of these incredible creatures created on day 5 and day 6 of the creation week.

Getting Started

This course has been developed to enhance learning about the diverse and amazing animals we see in our world. From the tiniest to the largest, you will be exploring what makes them unique as you celebrate their place in the world God created.

Although the material can be modified for your own educational purposes, either on an elementary or more advanced level, it has been organized in this PLP to fit a one-year course. Worksheets, simple sectional quizzes, and a final semester exam are all focused on *The World of Animals*. Other books noted in the calendar are primarily for added activities. The simple quizzes can be used in grading progress, assessing learning, or as test preparation for the final exam.

Worksheet Questions

The questions included in this parent lesson plan are designed for comprehension and can be used to evaluate the student's level of understanding while reading through *The World of Animals*. Answers to the questions can be given either verbally or written, based on the abilities of or the educational approach for the student and the requirements of your state. Younger students can do open-book tests if the parent/educator allows it.

Answer Key

An answer key for all questions and quizzes is available at the back of this parent lesson plan for your convenience.

Bonus Activities

Each section will have five optional bonus activities. These activities will each be worth 20 points and can include visual, reading, research, and writing activities. These are not required, but choosing one per section can be an additional learning experience.

SIMPLE ANIMALS:

- Choose an animal to write a one-page essay about.

- Create a short story using some of the animals in this section. Try to keep the story realistic in terms of how they live.

- Create a visual aquarium by drawing and coloring at least ten of the animals in this section. Take a small box, cut out the animals (if you have your parent's permission), and glue the animals on it to form your own paper aquarium.

- Take a map and identify five bodies of salt water and five bodies of fresh water.

- Take a small notebook to the grocery store and list some of the animals in this section that are found there.

WORMS, SNAILS, AND STARFISH

- Make a drawing that shows a worm on or under the ground or one in the water. Be sure to include other details of life that may surround it. Be creative — choose an unusual setting like a farm, flower-pot, park, forest, or beach.

- Write a short story that tells how a starfish tries to sneak up on its lunch.

- Check the ground after a rain or at night, if the soil is loose, to see if you can see any earthworms. Give a short report on what you find.

- Collect a worm or snail and try to identify it in the book — also make a list of some of its features that would help you identify it if you ever see it again.

- See how many empty snail shells you can find in your backyard. Are all of the snails the same size, color or shape? Write a short essay telling what you find.

INSECTS AND OTHER ANTHROPODS

- Set up an ant farm from a kit. Observe how the ants live, and write a one-page essay on what you observe or an interesting part of ant life.

- Draw three kinds of insects — try to be detailed and accurate in their size, shape, features, and colors. You can use the insects in this section as models for your drawing. Or make a drawing of a spider's web.

- See if you can find out if there are any unusual insects that are unique to your state or region. If you find one, write a short report and include a copy of an illustration or photo of the insect that you have found online (with your parent's permission) or from a reference book.

- Write a short story from the perspective of an insect — either celebrating its uniqueness or a challenge that its design has the ability to overcome (example: bumblebees can fly).

- Choose two insects and make a list of features from each, comparing and contrasting the two.

FISH

- Do you have a pet fish? If so, you can write a short story featuring your pet fish. Make the story an exciting adventure.

- If you have ever gone fishing, what types of fish have you caught? If you have never gone fishing, what type of fish do you enjoy eating (examples: tuna, tilapia, etc.)?

- What is the most amazing thing you have learned about fish in this section? Write a one-page essay telling what you found to be amazing.

- See if you can find any images of fish in a magazine. If you can, then see if you can identify them. Be sure to include whether they are freshwater or saltwater fish.

- Find a story book in the library or a movie that features a fish. Do a one-page report on the book or the movie.

AMPHIBIANS AND REPTILES

- If you have a favorite amphibian or reptile, what is it and why? Write the answers in a one-page essay.

- Do frogs live in your area? What kind are they, and where are they found? Do a short report on what you find.

- Why are sea turtles endangered, and what efforts are made to help protect them? This short essay will require additional research. Be sure to get your parent's permission to use the computer. You may also be able to find the answers in books or magazines at your local library.

- Find a map. Choose five snakes and mark up the map noting where each kind of snake can be found.

- Draw a specific kind of snake or a toad. Be sure to include details of its coloring and shape of its body.

BIRDS

- Take a few moments each day for a week and make note of all the different birds that you see.

- Are there any bird nests in your yard? Be careful not to disturb them, but make a weekly note on what you see happening in the nest. Include details on what kind of bird it may be, how many birds or eggs are in the nest, where the nest is in the yard, and how long it is before they take flight.

- Write a creative short story — include one bird from four different habitats (mountains, oceans, etc.) as characters in the story.

- Why are birds like parrots a big responsibility if you have one as a pet? Write a one-page essay that answers the question.

- How is a chicken different than a hawk? You can either make a drawing of each and point out the differences or write a list contrasting each bird.

MAMMALS — PART ONE: DOGS TO HORSES

- Do you have a dog as a pet? Write a two-page essay and include a drawing about your pet dog.

- See if you can find a book in the library about horses — either a reference book or a storybook. What is your favorite thing about a horse? Make a list of at least five things, and describe each one in detail.

- Give a one-page report on your favorite mammal. Tell why it is your favorite, where it is found, what makes it unique, etc.

- What is a fascinating fact about mammals that you have learned so far? Tell what it is and why it surprised you.

- Start a paper zoo in a blank notebook. Collect or create pictures of various kinds of animals; be sure to group them in some way (by kind, by area where they are found, by habitat, etc.).

MAMMALS — PART TWO: CATS TO MONKEYS

- Write a two-page play using several mammals in a specific location, like the ocean or the forest.

- See if you can list ten different kinds of cats.

- Mark up a map showing the locations of different types of whales.

- Often whales or dolphins get stuck on the shoreline in great numbers. Scientists don't really know why, but see if you can find any news stories about the efforts to save beached sea creatures. Get your parent's permission if you need to use the computer and get on the Internet.

- In the biblical account of creation, God makes all the animals and then He creates a man and a woman. How does that make us different from the animals? Draw a picture of what you think the Garden of Eden and its animals looked like.

First Semester Suggested Daily Schedule

Date	Day	Assignment	Due Date	✓	Grade
		First Semester-First Quarter — *The World of Animals, Dinosaur Activity Book*			
Week 1	Day 1	A World Full of Animals: Introduction • Read Pages 6-9 *The World of Animals* • (WA) • **Worksheet 1** • Pages 19-20 • (LP)			
	Day 2				
	Day 3	**Section 1** • Simple Animals: The Micro-World of Protists, Floating Protists • Read Pages 10-15 • (WA) **Worksheet 1** • Pages 21-22 • Lesson Planner • (LP)			
	Day 4				
	Day 5	Read and Color Pages 2-5 • *Dinosaur Activity Book* • (DAB)			
Week 2	Day 6	**Section 1** • Simple Animals: Water Bears and Wheel Animals, Sponges • Read Pages 16-19 • (WA) **Worksheet 2** • Pages 23-24 • (LP)			
	Day 7				
	Day 8	**Section 1** • Simple Animals: Comb Jellies, The "Nettle Animals" Read Pages 20-25 • (WA) • **Worksheet 3** • Pages 25-26 • (LP)			
	Day 9				
	Day 10	Read and Color Pages 6-9 • Complete Puzzle • (DAB)			
Week 3	Day 11	**Section 1** • Simple Animals: Jellyfish, More Jellyfish Read Pages 26-29 • (WA) • **Worksheet 4** • Pages 27-28 • (LP)			
	Day 12				
	Day 13	**Section 1** • Hydroids, Sea Anemones Read Pages 30-33 • (WA) • **Worksheet 5** • Pages 29-30 • (LP)			
	Day 14				
	Day 15	Read and Color Pages 10-13 • Complete Activity • (DAB)			
Week 4	Day 16	**Section 1** • Anemones and Partners, Corals, Coral Reefs Read Pages 34-39 • (WA) • **Worksheet 6** • Pages 31-32 • (LP)			
	Day 17	Study Day			
	Day 18	**Section 1 Simple Animals: Quiz 1** • Pages 137-138 • (LP)			
	Day 19				
	Day 20	Read and Color Pages 14-17 • Complete Activities • (DAB)			
Week 5	Day 21	**Section 2** • Worms, Snails, & Starfish: Earthworms and Leeches, Bristleworms • Read Pages 40-45 • (WA) **Worksheet 1** • Pages 33-34 • (LP)			
	Day 22				
	Day 23	**Section 2** • Worms, Snails, & Starfish: Flatworms, Flukes and Tapeworms • Read Pages 46-49 • (WA) **Worksheet 2** • Pages 35-36 • (LP)			
	Day 24				
	Day 25	Read and Color Pages 18-21 • Start Flip-Flap Fossil Reconstruction Fun Activity • (DAB)			

Date	Day	Assignment	Due Date	✓	Grade
Week 6	Day 26	**Section 2** • Worms, Snails, & Starfish: Roundworms, Other Worms • Read Pages 50-53 • (WA) **Worksheet 3** • Pages 37-38 • (LP)			
	Day 27				
	Day 28	**Section 2** • Worms, Snails, & Starfish: The Range of Mollusks, Slugs and Snails • Read Pages 54-57 • (WA) **Worksheet 4** • Pages 39-40 • (LP)			
	Day 29				
	Day 30	Read and Color Pages 22-25 • Continue Flip-Flap Fossil Reconstruction Fun Activity • (DAB)			
Week 7	Day 31	**Section 2** • Worms, Snails, & Starfish: Sea Snails and Limpets, Seashells • Read Pages 58-61 • (WA) **Worksheet 5** • Pages 41-42 • (LP)			
	Day 32				
	Day 33	**Section 2** • Worms, Snails, & Starfish: Bivalve Mollusks, Squid, Octopuses, and Cuttlefish • Read Pages 62-65 • (WA) **Worksheet 6** • Pages 43-44 • (LP)			
	Day 34				
	Day 35	Read and Color Pages 26-29 • Complete Flip-Flap Fossil Reconstruction Fun Activity • (DAB)			
Week 8	Day 36	**Section 2** • Worms, Snails, & Starfish: Starfish and Seastars, Sea Urchins and Sea Cucumbers • Read Pages 66-69 • (WA) **Worksheet 7** • Pages 45-46 • (LP)			
	Day 37	Study Day			
	Day 38	**Section 2 Worms, Snails, & Starfish: Quiz 2** Pages 139-140 • (LP)			
	Day 39				
	Day 40	Read and Color Pages 30-32 • Complete Maze • (DAB)			
Week 9	Day 41	**Section 3** • Insects & Other Arthropods: The World of Bugs, Beetles and Weevils • Read Pages 70-75 • (WA) **Worksheet 1** • Pages 47-48 • (LP)			
	Day 42				
	Day 43	**Section 3** • Insects & Other Arthropods: Butterflies and Moths, Bees, Wasps, and Ants • Read Pages 76-79 • (WA) **Worksheet 2** • Pages 49-50 • (LP)			
	Day 44				
	Day 45	Read and Color Pages 33-36 • Complete *Protoceratops* Toy (DAB)			
First Semester-Second Quarter — *The World of Animals, Dinosaur Activity Book The Complete Aquarium Adventure*					
Week 1	Day 46	**Section 3** • Insects & Other Arthropods: Flies, Dragonflies and Damselflies • Read Pages 80-83 • (WA) **Worksheet 3** • Pages 51-52 • (LP)			
	Day 47				
	Day 48	**Section 3** • Insects & Other Arthropods: Bugs, Crickets and Grasshoppers • Read Pages 84-87 • (WA) **Worksheet 4** • Pages 53-54 • (LP)			
	Day 49				
	Day 50	Read and Color Pages 37-39 • Complete Activities • (DAB)			

Date	Day	Assignment	Due Date	✓	Grade
Week 2	Day 51	**Section 3** • Insects & Other Arthropods: Barnacles and Other Crustaceans, Spiders • Read Pages 88-91 • (WA) **Worksheet 5** • Pages 55-56 • (LP)			
	Day 52				
	Day 53	**Section 3** • Insects & Other Arthropods: Fleas, Lice, and Other Insects, Crabs, Lobsters, and Shrimps Read Pages 92-95 • (WA) • **Worksheet 6** • Pages 57-58 • (LP)			
	Day 54				
	Day 55	Read and Color Pages 40-43 • Complete Activity • (DAB)			
Week 3	Day 56	**Section 3** • Insects & Other Arthropods: Scorpions and Other Arachnids, Centipedes and Millipedes Read Pages 96-99 • (WA) • **Worksheet 7** • Pages 59-60 • (LP)			
	Day 57	Study Day			
	Day 58	**Section 3 Insects & Other Arthropod: Quiz 3** Pages 141-142 • (LP)			
	Day 59				
	Day 60	Read and Color Pages 44-47 • (DAB)			
Week 4	Day 61	**Section 4** • Fish: What Are Fish?, Sharks and Rays Read Pages 100-105 • (WA) • **Worksheet 1** • Pages 61-62 • (LP)			
	Day 62				
	Day 63	**Section 4** • Fish: Sturgeons and Gars, Eels and Herrings Read Pages 106-109 • (WA) • **Worksheet 2** • Pages 63-64 • (LP)			
	Day 64				
	Day 65	Read and Color Pages 48-52 • Complete Activities • (DAB)			
Week 5	Day 66	**Section 4** • Fish: Salmon, Pike, and Hatchetfish, Characins, Carp, and Catfish • Read Pages 110-113 • (WA) **Worksheet 3** • Pages 65-66 • (LP)			
	Day 67				
	Day 68	**Section 4** • Fish: Cod, Anglerfish, and Toadfish, Scorpionfish and Seahorses • Read Pages 114-117 • (WA) **Worksheet 4** • Pages 67-68 • (LP)			
	Day 69				
	Day 70	Read and Color Pages 53-58 • Complete Activities • (DAB)			
Week 6	Day 71	**Section 4** • Fish: Flyingfish, Silversides, and Killifish, Flatfish and Triggerfish • Read Pages 118-121 • (WA) **Worksheet 5** • Pages 69-70 • (LP)			
	Day 72				
	Day 73	**Section 4** • Fish: Perch, Groupers, and Drums, Tunas and Marlins • Read Pages 122-125 • (WA) **Worksheet 6** • Pages 71-72 • (LP)			
	Day 74				
	Day 75	Read and Color Pages 59-63 • Complete Activities • (DAB)			

Date	Day	Assignment	Due Date	✓	Grade
Week 7	Day 76	**Section 4** • Fish: Cichlids, Damsels, and Parrotfish, Blennies, Gobies, and Wrasses • Read Pages 126-129 • (WA) **Worksheet 7** • Pages 73-74 • (LP)			
	Day 77	Study Day			
	Day 78	**Section 4 Fish: Quiz 4** • Pages 143-144 • (LP)			
	Day 79				
	Day 80	Read and Color Pages 64-67 • Complete Activities • (DAB)			
Week 8	Day 81	**Aquarium Adventure**: Weigh Anchor (Follow Page 10 Outline): Devotion and Exploring the Depths 1 • Read Pages 12-13			
	Day 82	**Aquarium Adventure**: Weigh Anchor (Follow Page 10 Outline): Devotion and Exploring the Depths 2 • Read Pages 14-15			
	Day 83	**Aquarium Adventure**: Weigh Anchor (Follow Page 10 Outline): Devotion and Exploring the Depths 3 • Read Pages 16-17			
	Day 84	**Aquarium Adventure**: Weigh Anchor (Follow Page 10 Outline): Devotion and Exploring the Depths 4 • Read Pages 18-19			
	Day 85	**Aquarium Adventure**: Weigh Anchor (Follow Page 10 Outline): Devotion and Exploring the Depths 5 • Read Pages 20-21			
Week 9	Day 86	**Aquarium Adventure**: Weigh Anchor (Follow Page 10 Outline): Devotion and Exploring the Depths 6 • Read Pages 22-23			
	Day 87	**Aquarium Adventure**: Chart Your Course and Weigh Anchor (Follow Page 10 Outline): Devotion and Exploring the Depths 7 Read Pages 24-25			
	Day 88	**Aquarium Adventure: At the Aquarium • See Pages 29-34**			
	Day 89	**Aquarium Adventure:** All Ashore! • Page 112, After the Aquarium Begin Activities • Pages 113-148			
	Day 90	**Aquarium Adventure:** After the Aquarium Complete Activities Pages 113-148			
		Mid-Term Grade			

Second Semester Suggested Daily Schedule

Date	Day	Assignment	Due Date	✓	Grade
		Second Semester-Third Quarter — *The World of Animals, Dinosaur Activity Book*			
Week 1	Day 91	**Section 5** • Amphibians & Reptiles: What are Amphibians?, Newts and Salamanders • Read Pages 130-135 • (WA) **Worksheet 1** • Pages 75-76 • (LP)			
	Day 92				
	Day 93	**Section 5** • Amphibians & Reptiles: Frogs and Toads, What are Reptiles? • Read Pages 136-139 • (WA) **Worksheet 2** • Pages 77-78 • (LP)			
	Day 94				
	Day 95	Read and Color Pages 68-71 • (DAB)			
Week 2	Day 96	**Section 5** • Amphibians & Reptiles: Tortoises, Turtles, and Terrapins, Sea Turtles • Read Pages 140-143 • (WA) **Worksheet 3** • Pages 79-80 • (LP)			
	Day 97				
	Day 98	**Section 5** • Amphibians & Reptiles: Crocodiles and Alligators, How Crocodiles Breed • Read Pages 144-147 • (WA) **Worksheet 4** • Pages 81-82 • (LP)			
	Day 99				
	Day 100	Read and Color Pages 72-74 • Complete Activities • (DAB)			
Week 3	Day 101	**Section 5** • Amphibians & Reptiles: Iguanas, Agamids, and Chameleons, Geckos, Lacertids, and Teiid Lizards Read Pages 148-151 • (WA) • **Worksheet 5** • Pages 83-84 • (LP)			
	Day 102				
	Day 103	**Section 5** • Amphibians & Reptiles: Skinks, Monitors, and Slow Worms, Pythons, Boas, and Thread Snakes Read Pages 152-155 • (WA) • **Worksheet 6** • Pages 85-86 • (LP)			
	Day 104				
	Day 105	Read and Color Pages 75-79 • Complete Activities • (DAB)			
Week 4	Day 106	**Section 5** • Amphibians & Reptiles: Colubrid Snakes, Cobras, Vipers, and Rattlers • Read Pages 156-159 • (WA) **Worksheet 7** • Pages 87-88 • (LP)			
	Day 107	Study Day			
	Day 108	**Section 5 Amphibians & Reptiles: Quiz 5** Pages 145-146 • (LP)			
	Day 109				
	Day 110	Read and Color Pages 80-85 • Complete Jumping Jack • (DAB)			
Week 5	Day 111	**Section 6** • Birds: Flightless Birds, Seabirds Read Pages 160-165 • (WA) • **Worksheet 1** • Pages 89-90 • (LP)			
	Day 112				
	Day 113	**Section 6** • Birds: Shorebirds and Waterbirds, Herons, Ducks and Geese • Read Pages 166-169 • (WA) **Worksheet 2** • Pages 91-92 • (LP)			
	Day 114				
	Day 115	Read and Color Pages 86-89 • Complete Activities • (DAB)			

Date	Day	Assignment	Due Date	✓	Grade
Week 6	Day 116	**Section 6** • Birds: Birds of Prey, Gamebirds and Rails Read Pages 170-173 • (WA) • **Worksheet 3** • Pages 93-94 • (LP)			
	Day 117				
	Day 118	**Section 6** • Birds: Pigeons, Doves, and Parrots, Cuckoos and Turacos • Read Pages 174-177 • (WA) **Worksheet 4** • Pages 95-96 • (LP)			
	Day 119				
	Day 120	Read and Color Pages 90-93 • Complete Activities • (DAB)			
Week 7	Day 121	**Section 6** • Birds: Owls and Nightjars, Swifts and Woodpeckers • Read Pages 178-181 • (WA) **Worksheet 5** • Pages 97-98 • (LP)			
	Day 122				
	Day 123	**Section 6** • Birds: Crows, Shrikes, and Bowerbirds, Sparrows, Finches, and Weavers • Read Pages 182-185 • (WA) **Worksheet 6** • Pages 99-100 • (LP)			
	Day 124				
	Day 125	Read and Color Pages 94-97 • (DAB)			
Week 8	Day 126	**Section 6** • Birds: Warblers, Thrushes, and Flycatchers, Larks, Swallows, and Treecreepers • Read Pages 186-189 • (WA) **Worksheet 7** • Pages 101-102 • (LP)			
	Day 127	Study Day			
	Day 128	**Section 6 Birds: Quiz 6** • Pages 147-148 • (LP)			
	Day 129				
	Day 130	Read and Color Pages 98-100 • Complete Activities • (DAB)			
Week 9	Day 131	**Section 7** • Mammals: Egg-Laying Mammals, Marsupial Mammals • Read Pages 190-195 • (WA) **Worksheet 1** • Pages 103-104 • (LP)			
	Day 132				
	Day 133	**Section 7** • Mammals: Insect Eaters, Bats Read Pages 196-199 • (WA) **Worksheet 2** • Pages 105-106 • (LP)			
	Day 134				
	Day 135	Read and Color Pages 101-102 • Complete Activity • (DAB)			
Second Semester-Fourth Quarter — *The World of Animals, Dinosaur Activity Book The Complete Zoo Adventure*					
Week 1	Day 136	**Section 7** • Mammals: Anteaters, Sloths, and Armadillos, Rabbits, Hares, and Pikas • Read Pages 200-203 • (WA) **Worksheet 3** • Pages 107-108 • (LP)			
	Day 137				
	Day 138	**Section 7** • Mammals: Hyraxes and the Aardvark, Mice, Rats, and Cavies • Read Pages 204-207 • (WA) **Worksheet 4** • Pages 109-110 • (LP)			
	Day 139				
	Day 140	Read and Color Pages 103-109 • Complete Activities • (DAB)			

Date	Day	Assignment	Due Date	✓	Grade
Week 2	Day 141	**Section 7** • Mammals: Large Rodents, Squirrels and Chipmunks • Read Pages 208-211 • (WA) **Worksheet 5** • Pages 111-112 • (LP)			
	Day 142				
	Day 143	**Section 7** • Mammals: Deer, Camels, and Pigs, Antelopes, Wild Cattle, and Sheep • Read Pages 212-215 • (WA) **Worksheet 6** • Pages 113-114 • (LP)			
	Day 144				
	Day 145	Read and Color Pages 110-112 • Complete Activity • (DAB)			
Week 3	Day 146	**Section 7** • Mammals: Horses, Zebras, and Rhinos, Elephants Read Pages 216-219 • (WA) **Worksheet 7** • Pages 115-116 • (LP)			
	Day 147	Study Day			
	Day 148	**Section 7 Mammals, Part 1; Quiz 7, Part 1** Pages 149-150 • (LP)			
	Day 149	**Section 7** • Mammals: Cats, Dogs, Foxes, and Hyenas Read Pages 220-223 • (WA) **Worksheet 8** • Pages 117-118 • (LP)			
	Day 150	Read and Color Pages 113-116 • Complete Activities • (DAB)			
Week 4	Day 151	**Section 7** • Mammals: Bears, Raccoons, and Pandas, Weasels, Mongooses, and Civets • Read Pages 224-227 • (WA) **Worksheet 9** • Pages 119-120 • (LP)			
	Day 152				
	Day 153	**Section 7** • Mammals: Seals, Sea Lions, and Sea Cows Read Pages 228-229 • (WA) **Worksheet 10** • Pages 121-122 • (LP)			
	Day 154				
	Day 155	Read and Color Pages 117-119 • Complete Crossword Puzzle (DAB)			
Week 5	Day 156	**Section 7** • Mammals: Great Whales Read Pages 230-231 • (WA) **Worksheet 11** • Pages123-124 • (LP)			
	Day 157				
	Day 158	**Section 7** • Mammals: Dolphins and Porpoises Read Pages 232-233 • (WA) **Worksheet 12** • Pages 125-126 • (LP)			
	Day 159				
	Day 160	Read and Color Pages 120-121 • (DAB)			
Week 6	Day 161	**Section 7** • Mammals: Lemurs and Bushbabies, African Monkeys • Read Pages 234-237 • (WA) **Worksheet 13** • Pages 127-128 • (LP)			
	Day 162	Study Day			
	Day 163	**Section 7 Mammals, Part 2: Quiz 7, Part 2** Pages 151-152 • (LP)			
	Day 164				
	Day 165	Read and Color Pages 122-123 • (DAB)			

Date	Day	Assignment	Due Date	✓	Grade
Week 7	Day 166	**Section 7** • Mammals: Asian and American Monkeys Read Pages 238-241 • (WA) **Worksheet 14** • Pages 129-130 • (LP)			
	Day 167	**Section 7** • Mammals: Gibbons, Orangs and Gorillas Read Pages 242-245 • (WA) **Worksheet 15** • Pages 131-132 • (LP)			
	Day 168	**Section 7** • Mammals: Chimpanzees Read Pages 246-247 • (WA) **Worksheet 16** • Pages 133-134 • (LP)			
	Day 169	Optional Semester Exam • Pages 153-156 • (LP)			
	Day 170	Read and Color Pages 124-125 • (DAB)			
Week 8	Day 171	**Zoo Adventure** Before the Zoo (Follow Page 10 Outline): Devotional and Looking Ahead 1 • Pages 12-13			
	Day 172	**Zoo Adventure** Before the Zoo (Follow Page 10 Outline): Devotional and Looking Ahead 2 • Pages 14-15			
	Day 173	**Zoo Adventure** Before the Zoo (Follow Page 10 Outline): Devotional and Looking Ahead 3 • Pages 16-17			
	Day 174	**Zoo Adventure** Before the Zoo (Follow Page 10 Outline): Devotional and Looking Ahead 4 • Pages 18-19			
	Day 175	**Zoo Adventure** Before the Zoo (Follow Page 10 Outline): Devotional and Looking Ahead 5 • Pages 20-21			
Week 9	Day 176	**Zoo Adventure** Before the Zoo (Follow Page 10 Outline): Devotional and Looking Ahead 6 • Pages 22-23			
	Day 177	**Zoo Adventure** Before the Zoo (Follow Page 10 Outline): Devotional and Looking Ahead 7 • Pages 24-25			
	Day 178	**Zoo Adventure: At the Zoo-Field Trip Day!** • See Pages 27-30			
	Day 179	**Zoo Adventure** Around the Campfire • Page 104 After the Zoo Begin Activities • Pages 105-140			
	Day 180	**Zoo Adventure:** After the Zoo Complete Activities Pages 105-140			
		Final Grade			

Animal Worksheets

for Use with

The World of Animals

Please read the assigned pages and then answer the following questions:

1. God could have made everything look the same, but we have incredible diversity among plants and animals. What does that say about the creativity of God?

2. What are the five kingdoms of living things?

3. How do plants and animals get their energy?

4. How are glacier grasshoppers and various desert creatures examples of animals designed to live in certain habitats?

5. When you are classifying animals, what are the two classifications used?

Please read the assigned pages and then answer the following questions:

1. What characteristics make an animal a simple one?

2. What is the name of the simplest life form?

3. Sea anemones are close relatives of what creatures?

4. Why are many protists considered parasites?

5. How is malaria, caused by protists called plasmodias, spread?

6. What connection do floating protists have to plankton?

7. How does much of plankton make its own food?

8. Why are some of the simple animals also called "producers"?

9. What is a diatom, and where are they found?

10. What causes a "red" tide?

Please read the assigned pages and then answer the following questions:

1. Why are the bryozoan called "moss" animals?

2. How does a moss animal protect itself?

3. Do all moss animals live in the ocean? If not, what other places?

4. How does sea moss spread to new areas?

5. Where do sea mats like to grow?

6. What is another name for water bears?

7. What do water bears eat?

8. What is another name for rotifers?

9. Describe the physical appearance of hairybacks.

10. Where do water bears live?

Please read the assigned pages and then answer the following questions:

1. How do sponges eat?

2. How do sponges protect themselves?

3. How many cells make up the body of placozoans?

4. How do sponges serve as "living filters"?

5. How do comb jellies move through the water?

6. What is bioluminescence?

7. What are three physical characteristics of comb jellies?

8. What are two other names for nettle animals?

9. What are the two basic body shapes of nettle animals?

10. What are the stinging cells on tentacles on the "nettle animals" used for?

Please read the assigned pages and then answer the following questions:

1. How large are jellyfish?

2. How do jellyfish swim?

3. What animals eat jellyfish?

4. Why do many jellyfish die after being washed ashore?

5. How do jellyfish see?

6. Give one example of a colonial cnidarian.

7. Why are box jellyfish harmful to swimmers or people wading in the ocean?

8. What are siphonophores?

9. How are box jellyfish different than true jellyfish?

10. In what country's waters does the southern sea wasp live?

11. What unique characteristic does the by-the-wind-sailor have?

Please read the assigned pages and then answer the following questions:

1. What do most hydroids resemble?

2. What is the purpose of each type of polyp inside an obelia colony?

3. Where do sea firs grow?

4. Where do hydras live?

5. Which colonial hydroids resemble true coral?

6. What two types of anthozoans are called flower-like animals?

7. Do sea anemones have a medusa?

8. What colors can be found on a beadlet anemone?

9. How do sea anemones feed?

10. Where is the mouth of an anemone found?

Please read the assigned pages and then answer the following questions:

1. What are four other animals that anemones form partnerships with?

2. Define the word *symbiosis*.

3. What is the partnership between hermit crabs and anemones?

4. What protects clown fish from their anemone partner?

5. Describe the protective cases of soft corals.

6. How do stony coral polyps protect themselves from danger?

7. What do feeding polyps rely on to bring them food?

8. What are the three types of coral?

9. What is a coral reef, and how are they formed?

10. Why are coral reefs one of the richest wildlife habitats on earth?

11. What are three prevalent threats to coral reefs and coral animals?

Please read the assigned pages and then answer the following questions:

1. How many different groups of worm-like creatures are there?

2. Where do most worms live?

3. How do worms absorb oxygen?

4. Why do worms need the protection of the places in which they live?

5. How do earthworms help to keep the soil fertile?

6. The world's largest earthworms are giant Gippsland earthworms. Where are they found, and how large can they grow?

7. Where do sludge worms live, and what other name does their red color inspire?

8. Are leeches a type of worm? If so, what type and what do they eat?

9. What are two types of bristleworms?

10. How do peacock worms protect themselves from predators?

Please read the assigned pages and then answer the following questions:

. What is the name for the simplest of all worms, and how does their body differ from annelid worms like earthworms?

. Where do flatworms live?

. When are freshwater planarians (flatworms that are not parasites) most active in ponds and streams?

. What two types flatworms are parasites?

. Where do flukes live?

6. What purpose does the slimy mucus have on a fluke's body?

7. What ancient people also suffered from parasites like tapeworms that can grow very long?

8. Why are flukes said to have "a complicated life"?

9. How do tapeworms attach themselves to the gut wall of the animal it is living within?

10. What should you do with food to make sure that any possible tapeworm eggs or larvae or killed?

Please read the assigned pages and then answer the following questions:

1. True or false: Roundworms are very rare and hard to find.

2. True or False: Most roundworms are too small to be seen.

3. True or false: Roundworms help balance nature by feeding on dead and decaying animals.

4. True or false: A shovel can contain half a billion round worms.

5. True or false: People can die from hookworms getting into their bodies and remaining there untreated.

6. Name three other types of worms that are not segmented worms, flatworms, or roundworms.

7. What is the name of worms that live in groups and look like a mass of living hair?

8. Nemerteans live in the sea and are also known by what other name?

9. Beard worms have no stomach or intestines; how do they eat?

10. Where do tongue worms live?

Please read the assigned pages and then answer the following questions:

1. Snails, slugs, whelks, cockles, clams, octopuses, squid, cuttlefish, and mussels are all known by what name?

2. What two groups of mollusks live in fresh water?

3. What is the large, cloak-like part of a mollusk's main body called?

4. Of the seven main groups of mollusks, what are the three most familiar ones?

5. What is a mollusk's foot used for?

6. What is a mollusk's radula?

7. Do gastropods have hard or soft bodies?

8. Some land and sea snails have a flap called an operculum — what is its purpose?

9. How long does it take baby snails to reach maturity?

10. What are the creatures called that look like snails that have lost their shell, though some have a shell under part of their skin?

Please read the assigned pages and then answer the following questions:

1. How many more types of snails live in the oceans than live on land or in fresh water combined?

2. What type of shell may have been used as an early type of money?

3. How does the coloring of land snails vary as compared to sea snails?

4. What do snails use to breathe?

5. What do limpets eat?

6. Describe a sea hare.

7. Where do abalones live, and can you eat them?

8. Why are winkles or periwinkles among the most common sea snails?

9. Describe your favorite type of sea shell and what purpose a shell serves.

10. Name three different groups of shells.

Please read the assigned pages and then answer the following questions:

1. What does bi-valve mean?

2. How can scallops escape danger?

3. How do most bivalve mollusks eat?

4. Some bivalves can burrow holes in what places?

5. How does a pearl form?

6. What does the word *cephalopod* mean?

7. What are octopus tentacles covered in to hold their prey and move over rocks?

8. Cuttlefish are known as "chameleons of the sea" for what reason?

9. How do the cuttlefish control the coloration of their skin?

10. How is the nautilus unique among the cephalopods?

Please read the assigned pages and then answer the following questions:

1. Echinoderms include what groups of animals?

2. The word *echinoderm* means what?

3. Where do echinoderms live?

4. Which starfish likes to eat coral?

5. What do most starfish eat?

6. How do starfish overcome the challenge of shells when eating clams and bivalve mollusks?

7. How does the brittlestar live up to its name?

8. What creatures are described by the word *echinoderm*?

9. Describe what sea cucumbers look like.

10. Sea squirts are tunicates rather than echinoderms. How did they get their name?

Please read the assigned pages and then answer the following questions:

1. Of all the animals in the world, eight of every ten are what?

2. What three key features help insects to be so diverse, widespread, and successful?

3. How many legs do arachnids have?

4. Animals with jointed legs are part of a group of animals known as what?

5. What purpose does the exoskeleton or cuticle serve on an arthropod?

6. Name two kinds of arthropods.

7. Name the three sections of an insect's body.

8. Beetles and weevils make up the largest sub-group of insects known as what?

9. Why is the coloring of a ladybug a warning to possible predators?

10. What two things help tiger and ground beetles to be active and effective hunters?

11. The kind of beetle that often lights up the night sky is commonly known as what?

12. What do weevils eat?

13. How do diving beetles row themselves through the water?

14. The kind of beetle that can be seen swirling around on the top of the water is known as what?

15. The Goliath beetle is the world's heaviest insect. Where is it found?

Please read the assigned pages and then answer the following questions:

1. What is the second largest sub-group of insects?

2. What does the word *lepidoptera* mean?

3. How are butterflies and moths different?

4. Describe the life cycle of a butterfly or moth that is known as a complete metamorphosis.

5. What type of butterfly takes the long trip from Canada to Mexico each year in late summer?

6. The third largest sub-group of insects are what?

7. Which type of ant marches through the tropical forests in columns?

8. How is the tiny fairy fly misnamed?

9. How many wings do bees and wasps have?

10. Out of all the insects, which kind has the largest and most complicated colonies?

Please read the assigned pages and then answer the following questions:

1. What is the main feature of flies, and how does this help them to move?

2. Why are two-winged or true flies like hoverflies called the "supreme acrobats of the insect world"?

3. Do flies eat the same thing throughout every stage of their life cycle?

4. What is the disease that is caused by the tsetse fly in Africa?

5. What do mosquitoes, gnats, and midges need to need to eat to help their eggs develop?

6. Damselfly and dragonfly larvae have what in common?

7. What do dragonfly larvae eat?

8. How many times can a mayfly larva shed its skin?

9. How long is the life span of an adult stone fly?

10. Caddisfly larva will sometimes use empty snail shells for what?

Please read the assigned pages and then answer the following questions:

1. True or false: The word *bugs* is the correct one for any type of insect.

2. True or false: A Pondskater lives underwater.

3. True or false: The long tail of a water scorpion is actually a snorkel for them to breathe.

4. True or false: Some cicada nymphs will stay underground for 20 years before emerging, shedding their skin to become adults.

5. True or false: The casing or exoskeleton is thin between the sections of an arthropod's leg to allow it to bend.

6. True or false: Cockroaches can become serious pests.

7. True or false: Parental behavior, or caring, is very common among insects.

8. True or false: Crickets and grasshoppers belong to the subgroup called Orthopterans.

9. True or false: Crickets and grasshoppers sing to warn others away or to find a mate.

10. True or false: Stick and leaf insects are also known as phasmids.

11. True or false: Stick and leaf bugs camouflage themselves to hide from predators.

12. True or false: Mantids are not fierce hunters.

13. True or false: The praying mantis looks like it is praying because it is waiting to pounce on its prey.

14. True or false: Some of the longest insects are stick insects.

15. True or false: The larva of grasshoppers and crickets are called nymphs.

Please read the assigned pages and then answer the following questions:

1. The animals that fleas and lice live and feed upon are called what?

2. How far can an average flea jump?

3. How do fleas escape danger?

4. Describe a silverfish.

5. What do thrips eat, and what is their other name?

6. What are the most common animals on the land?

7. What group of arthropods occupy the salty waters of the world's oceans?

8. Why do hermit crabs have to "move house"?

9. How many pairs of antennae does a lobster have?

10. What is the difference in how shrimps and prawns move?

Please read the assigned pages and then answer the following questions:

1. True or false: Barnacles never attach themselves to ships or to other creatures.

2. True or false: Fairy shrimps swim upside-down and have no defense against predators.

3. True or false: Crustaceans cannot live along seashores or in the ocean.

4. True or false: Daphnia is another name for the common water flea.

5. True or false: Ostracods look like mini versions of mussels or clams.

6. True or false: Brine shrimp cannot live in warm or salty waters.

7. True or false: Spiders are part of the group of arthopods known as arachnids.

8. True or false: No spiders live in or on water.

9. True or false: Spiders can make webs using a part of their body called spinnerets.

10. True or false: Wolf spiders don't make webs and all spiders have poisonous fangs.

Please read the assigned pages and then answer the following questions:

1. What is the stinger on the end of a scorpion's tail used for?

2. What are the animals known as "false scorpions" called?

3. What are the two smallest arachnids?

4. How is a harvest spider different than a true spider?

5. What are two diseases that are spread by ticks?

6. What does the name *centipede* mean?

7. How many legs do most centipedes have?

8. What does the name *millipede* mean?

9. How many legs do most millipedes have?

10. Give two ways that millipedes and centipedes are different.

Please read the assigned pages and then answer the following questions:

1. Name a kind of fish that can live out of the water and breathe air.

2. What unique method do angler fish use to catch their food?

3. What is a vertebrate, and are fish vertebrates?

4. Are fish cold- or warm-blooded?

5. What unique feature is the reason some fish are called "cartilaginous" fish?

6. Is a dogfish a fish or a shark?

7. Are sharks covered in scales like a fish?

8. All chimeras have this specific feature for defense — what is it?

9. Sharks and rays all produce their young in the same way — yes or no?

10. What is the largest fish in the world?

Please read the assigned pages and then answer the following questions:

1. True or false: The paddlefish swims with its mouth open.

2. True or false: Scutes on a young sturgeon are scales.

3. True or false: The pirarucu or arapaima is a huge bonytongue fish that lives in the Amazon region of South America.

4. True or false: Elephant-nose fish are electric!

5. True or false: Most eels have no pelvic (rear side) fins.

6. True or false: Herrings eat other small fish and plankton but are not preyed upon by other creatures.

7. True or false: Garden eels live in colonies.

8. True or false: Moray eels live in freshwater lakes.

9. True or false: Most types of herring choose to lay their eggs on the seabed.

10. True or false: There are no eels in the deeper parts of the oceans.

Please read the assigned pages and then answer the following questions:

1. Where do salmon and trout live?

2. What are the sharp teeth of these fish used for?

3. What are the two forms of the common trout?

4. What is the largest member of the pike group?

5. How do pike capture their food?

6. The characins, carp, and catfish form what portion of all fish?

7. Do carp have teeth?

8. A piranha is which kind of fish: salmon, pike, hatchetfish, characins, carp, or catfish?

9. What feature helped give catfish their name?

10. What unique feature do knifefish have?

Please read the assigned pages and then answer the following questions:

1. Name a type of cod fish.

2. How does the shape of a fish's body impact the way it finds food?

3. Why are toadfish named toadfish?

4. How do frogfish hide from their prey?

5. What shapes are scorpionfish found to have?

6. Are all scorpionfish predators and, if so, what do they eat?

7. What is a common feature of the pipefish and the seahorse groups?

8. How do sticklebacks get their name?

9. What is unusual about how the seahorse reproduces?

10. Why are millions of seahorses killed each year?

Please read the assigned pages and then answer the following questions:

. Needlefish and longtoms are part of what group of fish?

. Do flyingfish fly?

. What feature gives the halfbeak its name?

. Where are grunions found?

. Guppies belong to which fish group?

6. What do guppies eat?

7. Do guppies lay eggs?

8. What is the only region of the world that you will not find flatfish?

9. What is unique about the triggerfish group?

10. What is the purpose of the porcupinefish being able to puff itself up, and how does that affect its many spines?

Please read the assigned pages and then answer the following questions:

. Almost half of all kinds of fish are part of this group — what is it?

. How do groupers hunt?

. What two perch-like fish make noises that helped to give them their names?

. Perch make up the largest group of perch-like fish — around how many species are there?

. Can some groupers grow even bigger than a human being?

6. What do walleye fish eat when they are newly hatched?

7. Scombrids include what type of fish?

8. Marlins and sailfish are also called what?

9. Are large barracudas a threat to human divers?

10. What are three possible purposes for the swordfish's "sword"?

Please read the assigned pages and then answer the following questions:

. Cichlids primarily live in what type of water?

. Where are the teeth of a parrotfish located?

. What type of fish has formed a partnership with sea anemones?

. Where are blennies found?

. Do gobies live alone or in schools?

6. What is unusual about a blennie's eyes?

7. Why are wrasse called a "cleaner fish"?

8. What kind of goby fish can live outside of water for minutes or even hours?

9. Where are gobies found around the world?

10. What is the smallest fish?

Please read the assigned pages and then answer the following questions:

. True or false: Amphibians and reptiles have the same soft, moist skin.

. True or false: Both the poisonous snakes and constrictor snakes belong to the reptile group.

. True or false: Lizards can range in size from small geckos to the large Komodo dragon.

. True or false: Amphibians are said to have double lives because they begin in water and end up living on land as adults.

. True or false: Tadpoles are the larval stage of reptiles.

6. True or false: The three main kinds of amphibians are newts/salamanders, frogs/toads, and caecilians.

7. True or false: Frogs are the type of amphibian that has no legs.

8. True or false: The brighter the color of a frog the more horrible-tasting it is to predators.

9. True or false: Newts and salamanders like to hunt at night.

10. True or false: A salamander can lay as few eggs as 4 or 5 and as many as 5,000, depending on the species

Please read the assigned pages and then answer the following questions:

. What animals make up the biggest group of amphibians?

. A frog can leap how many times the length of its own body?

. What do young tadpoles eat?

. Do all frogs make the same call, no matter what the species?

. If a frog is making a sound, is it probably a male or female?

. The bright colors of a poison-arrow frog mean what?

7. How does the flying frog move from tree to tree?

8. How do reptiles communicate?

9. Are caiman members of the crocodile group?

10. Why are terrapins efficient swimmers?

11. Why do reptiles depend on the heat of the sun to help regulate their body temperature?

12. What is torpor?

13. How is being cold-blooded an advantage for reptiles?

Please read the assigned pages and then answer the following questions:

1. What is the easiest way to recognize a turtle, terrapin, or tortoise?

2. Another name for this group of animals is what?

3. Do turtles, terrapins, and tortoises have teeth?

4. Are all chelonians predators?

5. What are the upper carapace and the lower plastron?

6. What is unusual about the camouflage of the matamata?

7. Where are the largest tortoises found?

8. What type of turtles make up the largest group within the chelonian group?

9. How is the shell of a sea turtle different than that of freshwater turtles?

10. Why is it so hard for newly hatched sea turtles to survive?

Please read the assigned pages and then answer the following questions:

. The three sub-groups of the crocodilians are what?

. What is the main difference between a crocodile and an alligator?

. When are crocodiles most active?

. How does the diet of a crocodile change with its size?

. How do crocodiles communicate?

6. Do female crocodiles take care of their eggs and young?

7. Crocodile eggs incubate for how long?

8. Where do crocodiles lay their eggs?

9. Where do alligators and caimans lay their eggs?

10. What factor impacts whether an alligator or crocodile egg will hatch a male or a female?

Please read the assigned pages and then answer the following questions:

1. What is the largest group of animals among reptiles?

2. Where are lizards most common?

3. Where do animals in this group live?

4. What does this group of animals eat?

5. What makes the marine iguana different than most other iguanas?

6. The flying dragon is what type of lizard?

7. What helps chameleons to hunt and to hide?

8. Do geckos have eyelids?

9. Name one kind of lacertid lizard.

10. Is a tuatara a lizard?

Please read the assigned pages and then answer the following questions:

1. What is the largest group of lizards?

2. What do Gila monsters eat?

3. Worm lizards are not lizards nor worms; they are part of a special group called what?

4. Do skinks have legs?

5. What is the world's largest lizard?

6. A goanna is another name for what species of lizard?

7. What are the two venomous lizards?

8. What is the only continent that does not have snakes?

9. Where do pipe snakes live?

10. How do pythons and boa kill their prey?

11. Are most snakes caring moms?

12. What is the name of the snake in swamps of South America that can grow to 30 feet long and can eat animals as large as goats?

Please read the assigned pages and then answer the following questions:

. What is the largest group of snakes?

. Are colubrids poisonous?

. Is the venom of the African boomslang strong enough to kill a human?

. What is the name for the venomous colubrid snakes based on their fangs?

. Do garter snakes lay eggs or give birth to live young?

6. What is the name of one snake that only eats eggs?

7. What are the two main groups of venomous snakes?

8. How do cobras warn off enemies?

9. How many species of sea snakes spend all their lives in the sea?

10. What do night-hunting pit vipers use to locate their prey?

Please read the assigned pages and then answer the following questions:

1. Do all birds use their wings for flying?

2. Which birds have the ability to swim?

3. Are birds warm- or cold-blooded?

4. What is one flightless bird that has become extinct?

5. What substance helps penguins to be waterproof?

6. How much do kiwi eggs weigh?

7. Ostriches and emus are examples of what type of flightless bird?

8. What is an example of a seabird that can spend weeks in the air?

9. How many species of gulls, skuas, and terns are there?

10. What bird resembles a penguin, with tight feathers to keep out water and the cold, and is plump?

11. Give three examples of flightless birds.

12. Give three examples of seabirds.

Please read the assigned pages and then answer the following questions:

1. Where do members of the Charadriiformes bird species live?

2. Where do these types of birds like to nest?

3. How are waders designed to help them find and eat food?

4. How much more can a pelican's chin- and throat-bag hold than its stomach?

5. This kind of bird has lobed rather than webbed feet; what is its name?

6. What does the oystercatcher use its bill for?

7. What are the largest waterfowl?

8. Spoonbills are close relatives of what birds?

9. What bird has one of the loudest bird calls?

10. Are geese and ducks powerful long-distance fliers?

Please read the assigned pages and then answer the following questions:

. Raptors, birds of prey, are known by these physical features.

. What helps birds of prey to be excellent hunters?

. What do hawks and eagles eat?

. What is the national bird of the United States, and what is the quality it represents?

. Give two examples of raptors.

6. What are the two main types of hawks?

7. What are gamebirds?

8. Name two types of gamebirds, and tell why they are difficult to hunt.

9. How many species of pheasants are there?

10. What bird is related to cranes?

Please read the assigned pages and then answer the following questions:

. What features do pigeons and parrots have in common?

. Parakeets are members of what bird group?

. Which birds are often used to carry messages because they are so good at finding their way home?

. What are the nests of doves and pigeons like?

. What type of bird has the ability to mimic the way we speak?

6. Where do macaws live?

7. Where do cuckoos like to lay their eggs?

8. How did the go-away bird get its name?

9. What do most turacos in Africa like to eat?

10. This type of ground cuckoo is well known in Mexico and the southwestern United States; it got its name for racing along roads in front of cars; what is it?

Please read the assigned pages and then answer the following questions:

. Which hunting birds hunt at night?

. How do an owl's very large eyes help it to hunt?

. An owl's ears are how many times more sensitive than those of a cat?

. What do nightjars eat?

. Which bird has been a symbol of wisdom since ancient times?

6. How many times does a hummingbird beat its wings every second?

7. Why are swifts rarely seen perching on trees?

8. Which birds are among the fastest birds and live up to their name?

9. How many species of hummingbirds are there and what do they eat?

10. What body feature do woodpeckers, toucans, barbets, and honeyguides have in common?

Please read the assigned pages and then answer the following questions:

. True or false: Another name for perching birds is passerines.

. True or false: Birds in the crow group are often very silent.

. True or false: Shrikes will impale prey on long thorns to save and eat later.

. True or false: Mockingbirds get their name from mimicking people's laughter.

. True or false: Bower birds like to decorate a place to impress a possible mate.

6. True or false: Weaver birds get their name from stealing loose threads.

7. True or false: A few birds have teeth.

8. True or false: Many species of sparrows, finches, and weaver birds migrate from Europe to Africa in winter.

9. True or false: Buntings is another name for a weaver bird.

10. True or false: The place where birds can temporarily store their meal when they have eaten is called a crop.

Please read the assigned pages and then answer the following questions:

1. What is the difference between a bird call and a bird song?

2. Why are so many warblers hard to see?

3. Where do many warblers live during winter?

4. Redstarts, wheatears, chats, and the European blackbird are all part of what bird group?

5. Which songbird can do song samples from over 100 other bird species?

6. Are larks, swallows, and treecreepers among the largest of the perching birds?

7. Where do swallows and house martins like to build their nests?

8. Why do some birds need thin, tweezer-like beaks?

9. Which birds like to live near mountain streams and can walk underwater to search for food?

10. What shape do swallows make their nests?

Please read the assigned pages and then answer the following questions:

. True or false: Marsupials are not a type of mammal.

. True or false: There are less than 400 species of mammals.

. True or false: The largest animal to ever live, the blue whale, was a mammal.

. True or false: Egg-laying mammals are also called monotremes.

. True or false: Echidnas use a long, sticky tongue to gather insects to eat.

6. True or false: Platypuses find their prey at night by detecting electrical signals from the muscles in their prey's body.

7. True or false: All mammals give birth to live young rather than lay eggs.

8. True or false: Marsupials are animals that have a pouch.

9. True or false: One type of kangaroo really lives in trees.

10. True or false: Marsupials are incapable of gliding.

Please read the assigned pages and then answer the following questions:

1. Insect-eating mammals are called what?

2. What is the word used to describe mammals that eat and are active at night?

3. Why is being warm-blooded a disadvantage for small mammals?

4. A fully grown European hedgehog has how many prickles covering its body?

5. How often do some small shrews have to eat or they will die?

6. Name three insectivores.

7. What percentage of mammal species are bats?

8. Are bats the only mammals that can truly fly?

9. What are the two main kinds of bats?

10. What do most bats use to locate their food?

11. What body feature do long-eared bats have that helps them find things with great accuracy?

12. What is the largest group of bats?

13. Do vampire bats really eat blood?

Please read the assigned pages and then answer the following questions:

. What are some of the characteristics of this group of mammals?

. What do anteaters eat?

. How many anteater species are there?

. How do armadillos defend themselves?

. Where did sloths get their name?

6. What is another mammal protected by a hard covering?

7. Why are rabbits and hares such good runners?

8. What is the difference between hares and rabbits?

9. What body feature helps rabbits and hares when eating?

10. What are leverets?

Please read the assigned pages and then answer the following questions:

. What is another name for hyraxes?

. Dassies are what type of hydrax?

. What is the only animal in the mammal group of tubulidents?

. Where do hydraxes make their homes?

. Describe an aardvark.

6. What is the largest mammal group?

7. Why is it not good that some mice and rats live near people?

8. What body feature of a hamster helps them store food?

9. Where do jerboas live?

10. What mice are known for taking very long winter sleeps?

Please read the assigned pages and then answer the following questions:

. True or false: The capybara of South America is the largest rodent of all.

. True or false: Beavers cut down trees just to eat on their soft, sap-rich bark.

. True or false: The den a beaver creates is called a lodge.

. True or false: There are three main groups of porcupines: the American, New World, and Old World.

. True or false: The cavy group includes chinchilla and viscachas.

6. True or false: Porcupines jab their spines into an attacker to protect themselves.

7. True or false: All squirrels live in trees.

8. True or false: Ground squirrels include marmots and prairie dogs.

9. True or false: There are many types of flying squirrels, most of them in Asia.

10. True or false: Forgetful squirrels are the reason some trees get planted.

Please read the assigned pages and then answer the following questions:

. Mammals with hooves rather than claws on their feet are called what name?

. Llamas, alpaca, guanaco, and vicuna are all close relatives of what animal?

. What are the two types of camels?

. What are among the most common and widespread of hoofed mammals?

. In what hooved mammals do even the females have horns?

6. Why do deer live in herds?

7. What mammals have horns but do not shed them like deer do?

8. Where does the yak live?

9. What body feature helps the musk ox to live in cold places?

10. What characteristics of antelopes and gazelles help them avoid predators?

Please read the assigned pages and then answer the following questions:

1. What hoofed mammals are included in the odd-toed ungulate group?

2. What do the animals in this group eat?

3. Where do tapirs live?

4. Zebras are also known by this descriptive name — what is it?

5. What are rhino horns made of?

6. What are the largest land animals?

7. Do elephants live alone or in small family groups?

8. How many hours may an elephant spend eating per day?

9. Can elephants drink water through their trunks?

10. An elephant's tusks are not horns, but instead are what?

Please read the assigned pages and then answer the following questions:

1. It is said that cats are a uniform group. What does this mean in terms of their body design?

2. What does the word *carnivore* mean?

3. What physical features are present in members of the Carnivora group?

4. Name three carnivores.

5. What makes the cats different from other carnivores?

6. Which cat is the fastest land animal over short distances?

7. What are the only cats that live in groups, and what are the groups called?

8. Where is the snow leopard found?

9. What is a feral cat?

10. What is the difference between dogs and cats in terms of body features?

11. What is the largest animal in the dog family?

12. How does the artic foxes' fur color help it survive?

13. How do dogs hunt differently than cats?

14. From which species do all domesticated dogs come?

15. A vixen is the female form of what member of the dog family?

Please read the assigned pages and then answer the following questions:

. Which is the rarest member of the bears, raccoons, and pandas group?

2. Where does the red panda live?

3. What body features help the polar bear survive in cold temperatures?

4. What are some of the things the common raccoon likes to eat?

5. What body feature helps the kinkajou move around in the trees?

6. What are the smallest of the mammal carnivores?

7. Which small carnivore in this group has a stinky solution to any threat?

8. Which animal is the largest in the weasel family?

9. Why are mongoose considered helpful?

10. Why is the otter successful as a swimmer?

Please read the assigned pages and then answer the following questions:

. Though they live in the water, where do seals, sea lions, and walruses prefer to stay?

. What do sea cows eat, and what are the other names for these creatures?

. Which seal will eat other seals as part of its diet?

. What do most seals and sea lions eat?

. What are baby seals called?

6. Why is the crabeater seal misnamed?

7. What are the two main groups of seals?

8. Why are seals at risk when they are on land?

9. Eared seals gather to rest on land at traditional sites called what?

10. What animal may have inspired legends of mermaids?

Please read the assigned pages and then answer the following questions:

. When do whales come ashore?

. How are great whales different than dolphins and porpoises?

. What almost drove whales to the point of extinction?

. What was the tool used to hunt whales called?

. What is the name for the group of whales that feed and travel together?

6. Which whale is known for its exceptionally long songs?

7. Baby whales are known as what?

8. The annual movement of whales from one place to another is known as what?

9. How deep can gray whales dive?

10. How do Blue and Humpback whales eat?

Please read the assigned pages and then answer the following questions:

. Toothed whales include what animals? Give two examples.

. What is the largest species of toothed whale, making it the largest carnivore on earth?

. What is an example of killer whales cooperating with one another to find prey?

. What does the sperm whale eat?

. Is the killer whale actually a whale or a dolphin?

6. How do dolphins communicate?

7. What is the distinctive feature of a narwhal?

8. Are dolphins found only in the ocean; if not, where?

9. What is an obvious difference in the body style of dolphins and porpoises?

10. What are the two kinds of white whales?

Please read the assigned pages and then answer the following questions:

. The group of animals that contains monkeys, apes, and lemurs is known as what?

. What are this group's primary body features that include a relatively large brain?

. Where is the only place you will find lemurs?

. What does the word *prosimians* mean?

. What are the primary body features of prosimians?

6. This animal is closely related to a baboon; what is it?

7. Which monkey is missing his thumb?

8. Do baboons live in the trees or on the ground?

9. Currently, what is one of the biggest threats they face?

10. How did howler monkeys get their name?

Please read the assigned pages and then answer the following questions:

. True or false: The largest group of Asian monkeys includes langurs.

. True or false: The Hanuman langur gets its water from the fruits it eats.

. True or false: Macaques are only found in North Africa.

. True or false: The proboscis monkey is named because of its ears.

. True or false: A douroucouli is another name for the owl monkey.

6. True or false: Howler monkeys live on fruits and young leaves.

7. True or false: The two main groups of New World or American monkeys are the small tamarins and marmosets, and the howler and wooly monkeys.

8. True or false: Squirrel monkeys live solitary lives.

9. True or false: The squirrel monkey is the most expert climber among monkeys.

10. True or false: American monkeys prefer to live in desert areas.

Please read the assigned pages and then answer the following questions:

. Are gibbons larger than monkeys?

. How many species of gibbon are there?

. When gibbons sing, does the male or female begin?

. Which species of gibbon sings by themselves and not as part of a pair?

. What is the average life span for a gibbon in the wild?

6. What are the four kinds of great apes?

7. What does the word *orangutan* mean?

8. Describe the body structure of a gorilla.

9. Among this group of animals, which one is the most vegetarian?

10. Which of the animals in this group are the largest tree-living mammals?

Please read the assigned pages and then answer the following questions:

. True or false: Chimps have the most complicated social lives.

. True or false: The two types of chimpanzees are the chimp and bonobo.

. True or false: A mother chimp will only care for its young for a couple of months.

. True or false: Chimps can live up to 50 years.

. True or false: Chimps are most active at night.

6. True or false: Chimps will not eat termites.

7. True or false: Illegal trapping is making chimp numbers fall rapidly.

8. True or false: At night, a chimp will make a nest platform from leaves and twigs to sleep on.

9. True or false: There are no basic differences between the common chimp and the pygmy chimp.

10. True or false: Chimps can make and use simple tools.

Quizzes & Test Section

Answer Questions: (5 Points Each Question)

. What are the five kingdoms of living things?

. How do plants and animals get their energy?

. When you are classifying animals, what are the two classifications used?

. What characteristics make an animal a simple one?

. What is the name of the simplest life form?

. Why are many protists considered parasites?

. How is malaria, caused by protists called plasmodias, spread?

. How does much of plankton make its own food?

. Why are some of the simple animals also called "producers"?

0. How does sea moss spread to new areas?

11. What is another name for rotifers?

12. How do sponges serve as "living filters"?

13. How do comb jellies move through the water?

14. What is bioluminescence?

15. What animals eat jellyfish?

16. How do jellyfish see?

17. What do most hydroids resemble?

18. How do sea anemones feed?

19. Define the word *symbiosis*.

20. What are the three types of coral?

Answer Questions: (5 Points Each Question)

. Where do most worms live?

. How do worms absorb oxygen?

. Are leeches a type of worm? If so, what type and what do they eat?

. What is the name for the simplest of all worms and how does their body differ from annelid worms like earthworms?

. What purpose does the slimy mucus have on a fluke's body?

. How do tapeworms attach themselves to the gut wall of the animal it is living within?

. Name three other types of worms that are not segmented worms, flatworms, or roundworms.

. Nemerteans live in the sea and are also known by what other name?

. Snails, slugs, whelks, cockles, clams, octopuses, squid, cuttlefish, and mussels are all known by what name?

0. What is a mollusk's foot used for?

11. Some land and sea snails have a flap called an operculum — what is its purpose?

12. How does the coloring of land snails vary as compared to sea snails?

13. What do snails use to breathe?

14. Name three different groups of shells.

15. What does bi-valve mean?

16. How does a pearl form?

17. Cuttlefish are known as "chameleons of the sea" for what reason?

18. The word *echinoderm* means what?

19. Where do echinoderms live?

20. What do most starfish eat?

Answer Questions: (5 Points Each Question)

. Of all the animals in the world, eight of every ten are what?

. What three key features help insects to be so diverse, widespread, and successful?

. How many legs do arachnids have?

. Animals with jointed legs are part of a group of animals known as what?

. Name two kinds of arthropods.

. How many wings do bees and wasps have?

. Out of all the insects, which kind has the largest and most complicated colonies?

. What is the main feature of flies, and how does this help them to move?

. Do flies eat the same thing throughout every stage of their life cycle?

0. True or false: The word *bugs* is the correct one for any type of insect.

11. True or false: Parental behavior, or caring, is very common among insects.

12. True or false: Stick and leaf bugs camouflage themselves to hide from predators.

13. The animals that fleas and lice live and feed upon are called what?

14. What group of arthropods occupy the salty waters of the world's oceans?

15. True or false: Barnacles never attach themselves to ships or to other creatures.

16. True or false: Crustaceans cannot live along seashores or in the ocean.

17. True or false: Spiders can make webs using a part of their body called spinnerets.

18. What is the stinger on the end of a scorpion's tail used for?

19. How is a harvest spider different than a true spider?

20. What does the name "centipede" mean?

Answer Questions: (5 Points Each Question)

1. Name a kind of fish that can live out of the water and breathe air.

2. What unique method do angler fish use to catch their food?

3. What is a vertebrate, and are fish vertebrates?

4. Are fish cold- or warm-blooded?

5. True or false: Most eels have no pelvic (rear side) fins.

6. True or false: Herrings eat other small fish and plankton but are not preyed upon by other creatures.

7. What are the sharp teeth of these fish used for?

8. What are the two forms of the common trout?

9. The characins, carp, and catfish form what portion of all fish?

10. What feature helped give catfish their name?

11. What is a common feature of the pipefish and the seahorse groups?

12. What is unusual about how the seahorse reproduces?

13. Needlefish and longtoms are part of what group of fish?

14. Guppies belong to which fish group?

15. Almost half of all kinds of fish are part of this group — what is it?

16. What do walleye fish eat when they are newly hatched?

17. Marlins and sailfish are also called what?

18. Cichlids primarily live in what type of water?

19. What type of fish has formed a partnership with sea anemones?

20. Why are wrasse called a "cleaner fish"?

| Q | *The World of Animals* Concepts & Comprehension | Quiz 5 | Scope: Section 5 | Total score: ____of 100 | Name |

Answer Questions: (4 Points Each Question)

. True or false: Amphibians and reptiles have the same soft, moist skin.

. True or false: Amphibians are said to have double lives because they begin in water and end up living on land as adults.

. True or false: The three main kinds of amphibians are newts/salamanders, frogs/toads, and caecilians.

. What animals make up the biggest group of amphibians?

. If a frog is making a sound, is it probably a male or female?

. How do reptiles communicate?

. Why do reptiles depend on the heat of the sun to help regulate their body temperature?

. What is torpor?

. What is the easiest way to recognize a turtle, terrapin, or tortoise?

0. What are the upper carapace and the lower plastron?

1. Why is it so hard for newly hatched sea turtles to survive?

2. What is the main difference between a crocodile and an alligator?

13. How does the diet of a crocodile change with its size?

14. What factor impacts whether an alligator or crocodile egg will hatch a male or a female?

15. What is the largest group of animals among reptiles?

16. What helps chameleons to hunt and to hide?

17. Is a tuatara a lizard?

18. What is the largest group of lizards?

19. What is the world's largest lizard?

20. What is the only continent that does not have snakes?

21. Are most snakes caring moms?

22. What is the largest group of snakes?

23. What is the name for the venomous colubrid snakes based on their fangs?

24. Do garter snakes lay eggs or give birth to live young?

25. What are the two main groups of venomous snakes?

Answer Questions: (5 Points Each Question)

. Do all birds use their wings for flying?

. Are birds warm- or cold-blooded?

. What is one flightless bird that has become extinct?

. How are waders designed to help them find and eat food?

. What are the largest waterfowl?

. Are geese and ducks powerful long-distance fliers?

. What helps birds of prey to be excellent hunters?

. What is the national bird of the United States, and what is the quality it represents?

. What are gamebirds?

0. Parakeets are members of what bird group?

11. What type of bird has the ability to mimic the way we speak?

12. Where do cuckoos like to lay their eggs?

13. How do an owl's very large eyes help it to hunt?

14. An owl's ears are how many times more sensitive than those of a cat?

15. How many times does a hummingbird beat its wings every second?

16. Which birds are among the fastest birds and live up to their name?

17. True or false: Birds in the crow group are often very silent.

18. True or false: A few birds have teeth.

19. True or false: The place where birds can temporarily store their meal when they have eaten is called a crop.

20. Where do swallows and house martins like to build their nests?

Answer Questions: (4 Points Each Question)

. True or false: There are fewer than 400 species of mammals.

. True or false: The largest animal to ever live, the blue whale, is a mammal.

. True or false: Egg-laying mammals are also called monotremes.

. True or false: Marsupials are animals that have a pouch.

. What is the word used to describe mammals that eat and are active at night?

. Why is being warm-blooded a disadvantage for small mammals?

. Are bats the only mammals that can truly fly?

. How do armadillos defend themselves?

. Where did sloths get their name?

0. What is the difference between hares and rabbits?

1. What is another name for hydraxes?

2. Why is it not good that some mice and rats live near people?

13. What body feature of a hamster helps them store food?

14. What mice are known for taking very long winter sleeps?

15. True or false: The den a beaver creates is called a lodge.

16. True or false: Porcupines jab their spines into an attacker to protect themselves.

17. True or false: All squirrels live in trees.

18. True or false: Forgetful squirrels are the reason some trees get planted.

19. Mammals with hooves rather than claws on their feet are called what name?

20. What are the two types of camels?

21. Why do deer live in herds?

22. What body feature helps the musk ox to live in cold places?

23. Zebras are also known by this descriptive name — what is it?

24. Do elephants live alone or in small family groups?

25. Can elephants drink water through their trunks?

Q	*The World of Animals*	Quiz 7	Scope:	Total score:	Name
	Concepts & Comprehension	Part 2	Section 7	____of 100	

Answer Questions: (4 Points Each Question)

. What does the word *carnivore* mean?

. What makes the cats different from other carnivores?

. What is a feral cat?

. From which species do all domesticated dogs come from?

. What body features help the polar bear to survive in cold temperatures?

. What are the smallest of the mammal carnivores?

. Why are mongoose considered helpful?

. What do sea cows eat, and what are the other names for these creatures?

. What are baby seals called?

0. What are the two main groups of seals?

1. Why are seals at risk when they are on land?

2. What almost drove whales to the point of extinction?

13. Which whale is known for its exceptionally long songs?

14. Baby whales are known as what?

15. How do Blue and Humpback whales eat?

16. What is the largest species of toothed whale, making it the largest carnivore on earth?

17. How do dolphins communicate?

18. What is an obvious difference in the body style of dolphins and porpoises?

19. The group of animals that contains monkeys, apes, and lemurs is known as what?

20. True or false: The two main groups of New World or American monkeys are the small tamarins and marmosets, and the howler and wooly monkeys.

21. True or false: The squirrel monkey is the most expert climber among monkeys.

22. Are gibbons larger than monkeys?

23. What does the word *orangutan* mean?

24. True or false: Chimps have the most complicated social lives.

25. True or false: At night, a chimp will make a nest platform from leaves and twigs to sleep on.

T	*The World of Animals* Concepts & Comprehension	Optional Final Exam	Scope: Animals	Total score: ____of 100	Name

Answer Questions: (2 Points Each Question)

. What are the five kingdoms of living things?

. How do plants and animals get their energy?

. When you are classifying animals, what are the two classifications used?

. What is the name of the simplest life form?

. Why are some of the simple animals also called "producers"?

. How do sponges serve as "living filters"?

. What is bioluminescence?

. Define the word *symbiosis*.

. Where do most worms live?

0. How do worms absorb oxygen?

1. Name three other types of worms that are not segmented worms, flatworms, or roundworms.

12. What do snails use to breathe?

13. What does bi-valve mean?

14. The word *echinoderm* means what?

15. What three key features help insects to be so diverse, widespread, and successful?

16. Do flies eat the same thing throughout every stage of their life cycle?

17. The animals that fleas and lice live and feed upon are called what?

18. What group of arthropods occupy the salty waters of the world's oceans?

19. How is a harvest spider different than a true spider?

20. What is a vertebrate, and are fish vertebrates?

21. Are fish cold- or warm-blooded?

22. What is a common feature of the pipefish and the seahorse groups?

23. Almost half of all kinds of fish are part of this group — what is it?

24. Cichlids primarily live in what type of water?

25. What animals make up the biggest group of amphibians?

26. Why do reptiles depend on the heat of the sun to help regulate their body temperature?

27. What is the main difference between a crocodile and an alligator?

28. How does the diet of a crocodile change with its size?

29. What helps chameleons to hunt and to hide?

30. What is the largest group of snakes?

31. Do garter snakes lay eggs or give birth to live young?

32. Do all birds use their wings for flying?

33. Are birds warm- or cold-blooded?

34. What helps birds of prey to be excellent hunters?

35. What type of bird has the ability to mimic the way we speak?

36. How do an owl's very large eyes help it to hunt?

37. True or false: A few birds have teeth.

38. True or false: Egg-laying mammals are also called monotremes.

39. True or false: Marsupials are animals that have a pouch.

40. What is the word used to describe mammals that eat and are active at night?

41. Why is being warm-blooded a disadvantage for small mammals?

42. Are bats the only mammals that can truly fly?

43. Mammals with hooves rather than claws on their feet are called what name?

44. What does the word *carnivore* mean?

45. What is a feral cat?

46. What body features help the polar bear to survive in cold temperatures?

47. What are the two main groups of seals?

48. Which whale is known for its exceptionally long songs?

49. How do dolphins communicate?

50. True or false: At night, a chimp will make a nest platform from leaves and twigs to sleep on.

Answer Keys

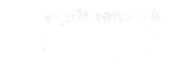

INTRODUCTION

Introduction – Worksheet 1

. He is powerful and creative, and created animals to survive in all climates and habitats.

. Monera, Protista, Plants, Fungi, and Animals

. Animals eat other living things; plants get their energy from sunlight.

. Glacier grasshoppers can be frozen alive, yet survive when they are thawed out; desert creatures can escape the heat by going underground or remaining dormant until the rains come.

. Vertebrates and invertebrates

SECTION 1: SIMPLE ANIMALS

Worksheet 1: The Micro-World of Protists; Floating Protists

. If it has very few basic body parts

. Sponges

. Jellyfish

. They thrive inside other living things, known as hosts.

. Through mosquito bites

. They form part of the plankton.

. It uses sunlight energy.

. They produce food for themselves rather than having to consume it.

. Plant-like protists with beautifully shaped, sculpted, and patterned shells

0. Huge numbers of dinoflagellates

Worksheet 2: Moss Animals; Water Bears and Wheel Animals

. They live in the sea and look like a patch of moss or doormat rather than an animal.

. They hide in their hard casing.

. No, in fresh water

4. They produce egg-like structures and are often blown to different places by the wind, starting new colonies.

5. On seaweed and kelp

6. Tardigrades

7. The sap of plants or on creatures smaller than itself

8. Wheel animals

9. They have round, bristly heads, a streamlined body, and forked tail.

10. Some live in the sea; some in fresh water

Worksheet 3: Sponges, Comb Jellies, The "Nettle Animals"

1. They filter tiny particles from the water.

2. Tiny spikes called spicules, unpleasant smell or taste

3. Around 1,000

4. They filter water, helping keep it clean and clear.

5. They drift or swim with the currents.

6. The ability of an animal to create its own light

7. Bands of comb-like cilia, two long sticky tentacles, soft transparent bodies

8. Cnidarians and coelenterates

9. The medusa and the polyp

10. Capture, paralyze, and kill prey

Worksheet 4: Jellyfish, More Jellyfish

1. Some are smaller than the tip of your finger; others are larger than a patio umbrella.

2. Contracting a ring of muscles to propel water out of the bell of the jellyfish and push the jellyfish through the water

3. Fish, squid, and sea turtles

4. Their bodies collapse and dry out.

5. Sense organs that can detect light or dark, as well as which way is up or down

6. Answers will vary; for examples: the man of war, by-the-wind-sailor

7. They have very powerful and painful stings; swimmers can be killed if they are stung by a group of them.

8. Colonies of polyps

9. Their body shape

10. Australia

11. A hard skeleton that acts like a mast to support a thin sail

Worksheet 5: Hydroids, Sea Anemones

1. Sea anemones

2. One kind does feeding, the other reproduction

3. On rocks along the shore below the low-tide line

4. Fresh water

5. Fire corals

6. Sea anemones and corals

7. No, they do not have this floating jellyfish-like stage.

8. Usually bright red, but can also be brown, orange, green

9. Trap fish and other small creatures in their tentacles

10. At the center of its tentacles

Worksheet 6: Anemones and Partners, Corals, Coral Reefs

1. Fish, shrimps, crabs, and worms

2. A helpful partnership between two different kinds of animals that is beneficial to both of them

3. Anemones live on the crab's shell, helping to protect it and giving it an opportunity to find more food.

4. A thick coating of slime

5. Spongy, rubbery jelly

6. They retreat into a protective cup.

7. The water's currents and waves

8. Stony, soft, gorgonian

9. A huge lump of rock made by millions of tiny animals called coral polyps.

10. They form a protected area from ocean waves, with plenty of places for creatures to call home, as well as seaweeds, fish, and other ocean creatures.

11. Pollution, mud and silt, crown-of-thorns starfish

SECTION 2: WORMS, SNAILS, & STARFISH

Worksheet 1: Earthworms and Leeches, Bristleworms

1. There are at least 30 different groups of worm-like creatures.

2. Most worms live only in water or under the ground in damp mud, sand, or soil.

3. Worms absorb oxygen through their thin, mois skin.

4. Worms are also soft, slow-moving, and vulnerable, so they gain some protection from living in tubes and tunnels.

5. Answers will vary — can include one or more o the following points: The earth passes through the worm's gut and nutritious bits like pieces of dead leaves are digested. The remains pass out o the worm as fine-grained droppings. Millions o earthworms keep soil fertile as they break down and recycle plant and animal remains. Their burrows allow air and moisture into the soil, fo plant roots to use, and to help water drain away

6. South Africa; over 3 feet long to more than 20 feet long

7. fresh water; bloodworm

8. Yes, they are like flattened types of worms. Some leeches eat smaller worms or insect grubs others eat blood from fish and even people.

9. Lugworms and ragworms

10. It retreats into the safety of its tube.

Worksheet 2: Flatworms, Flukes and Tapeworms

1. Flatworms; no ring-like sections or segments

- In water or as parasites in other animals

- Mainly at night

- Flukes and tapeworms

- Intestine, liver, heart, on or under the skin

- Protects it from the body fluids and digestive juices from the animal it lives within

- Egyptians

- Answers will vary — should focus around flukes possibly living in different animals at different times in its life cycle.

- With hooks and suckers

0. Cooking or curing the meat

Worksheet 3: Roundworms, Other Worms

- False

- True

- True

- False; half a million

- True

- Answers can include: ribbon worms, spoon worms, arrow worms, acorn worms, horsehair worms, peanut worms, beard worms, horseshoe worms, priapulids, tongue worms, spiny-headed worms

- Horsehair worms

- Ribbon worms

- Absorb nutrients through their body surface

0. Nose passages and lungs of mammals, birds, or reptiles

Worksheet 4: The Range of Mollusks; Slugs and Snails

- Mollusks

- Bivalves and snails

- Mantle

- Gastropods (snails, limpets, winkles), bivalves (mussels, clams, oysters, razorshells), and cephalopods (octopuses, cuttlefish, nautiluses, and squid)

- Answers can include: to catch prey, to cling to rocks, to slide around, to dig around in sand or mud.

6. Tongue

7. Soft

8. It is like a trapdoor that protects the snail in its shell.

9. Two years

10. Slugs

Worksheet 5: Sea Snails and Limpets; Seashells

1. Twice as many

2. Cowrie shells

3. Sea snails are often very colorful, while land snails are mostly gray and brown.

4. On land they have lungs; those in the sea have gills.

5. Seaweeds

6. Like a slug with a hare's long ears

7. Kelps beds in South Africa and California; yes, they are eaten by people and sea creatures like seals, sea otters, seabirds, sea lions

8. They are found all around the world, and are visible along the seashore.

9. Answers will vary; protection and as a home for the creature

10. Answers will vary but can include: tower shells, top shells, cowries, conches

Worksheet 6: Bivalve Mollusks; Squid, Octopuses, and Cuttlefish

1. With two valves

2. Swim away by opening and closing their shell and squirting out water to propel them

3. By letting the water bring them food and filtering out the edible particles

4. Wood or rock

5. A bit of grit or a parasite may enter the shell. If the mollusk cannot get rid of it, it is gradually covered in nacre, the substance that forms the smooth, white inner lining of a shell. This builds up in thin layers until the grit is completely

wrapped in a ball of shiny white nacre, forming a pearl.

6. Head-footed

7. Suckers

8. They are able to change both the color and patterns on their skin quickly, easily, and often

9. By their muscles and mixing different amounts of each color (yellow, red, brown, black)

10. It has an external shell like other mollusks; it is unable to squirt ink or easily escape danger

Worksheet 7: Starfish and Seastars; Sea Urchins and Sea Cucumbers

1. Starfish, seastars, feather stars, sea lilies, sea urchins, sand dollars, and sea cucumbers

2. Hedgehog skin

3. In the sea

4. The crown-of-thorns starfish

5. Mollusks, bivalves, crabs, worms, and other echinoderms

6. Answers will vary to include following processes: Starfish are very strong and can pry open the shells of clams and other bivalve mollusks. When attacking its victim, the starfish arches over it and clamps a couple of arms onto each side. It then pulls with great strength and stamina. As the two parts of the mollusk's shell begin to gape, the starfish turns its stomach inside out, through its mouth and into the gap in the shell. It then begins to digest the prey's flesh. Eventually, the shell opens and the starfish completes its meal.

7. Parts of its long, delicate arms often break off because they are brittle.

8. Hedgehog-like; sea urchins and sand dollars

9. Answers will vary and should include some of the following points: warty, sausage-shaped bodies, soft bodies, tough leathery skin, a mouth surrounded by tentacles, three rows of tube feet

10. If the tide goes out and leaves them exposed on the shore, they will often squirt out jets of water.

SECTION 3: INSECTS & OTHER ARTHROPODS

Worksheet 1: The World of Bugs; Beetles and Weevils

1. Insects

2. Hard outer body casing or exoskeleton, wings for flying, and six legs with flexible joints for running

3. Eight

4. Arthropods

5. Protection like a suit of armor

6. Answer will vary; may include: ants, beetles, bugs, cockroaches, flies, fleas, bees, termites, locusts, and other insects, spiders and scorpions, the multi-legged centipedes and millipedes, and crustaceans such as crabs, lobsters, and shrimps

7. head, thorax (chest), and abdomen

8. Coleopteran

9. It lets them know they taste horrible and to avoid them.

10. Large eyes to see prey, and long legs for catching it

11. A firefly

12. Seeds, fruits, and flowers

13. With their hairy legs

14. A whirligig beetle

15. Africa

Worksheet 2: Butterflies and Moths; Bees, Wasps, and Ants

1. Butterflies and moths

2. Scaly-winged

3. Most moths are active at night, most butterflies by day; moths are usually dull browns and grays, while many butterflies are brightly colored; most moths have feathery feelers while butterflies have thin, club-ended feelers

4. Answers will vary, but should include most of the following stages: laying eggs on a plant, eggs hatch as caterpillars, caterpillars grow and molt their skin several times before forming a chrysalis, a hard-case which surrounds it, inside

the chrysalis the body is being reformed into the various parts of an adult butterfly or moth before it emerges from the chrysalis

. Monarch or milkweed butterflies

. Bees, wasps, and ants

. Army ants

. They are not flies, but instead are tiny wasps.

. Four

0. Termites

Worksheet 3: Flies; Dragonflies and Damselflies

. One pair of proper wings, with a small second pair in the back which help them to balance

. Because they can hover in mid-air, fly sideways, and backward

. No

. Sleeping sickness

. Blood

. They live in fresh water for two years or more

. Tadpoles, small fish, and other small water creatures

. 25 times

. One or two weeks

0. A home

Worksheet 4: Bugs; Crickets and Grasshoppers

. False

. False

. True

. True

. True

. True

. False

. True

. True

0. True

1. True

2. False

3. True

14. True

15. True

Worksheet 5: Fleas, Lice, and Other Insects; Crabs, Lobsters, and Shrimps

1. Hosts

2. More than 12 inches

3. By leaping with its long, powerful back legs

4. Answers will vary, but can include no wings, but very long antennae and a three-forked tail, silvery color and bendy, fish-like movements.

5. The flowers of wheat, onions, carrots, fruits; thunderbugs or thunderflies

6. Insects

7. Crustaceans

8. When it gets too large for the abandoned sea snail shell it has taken over as its home

9. Two

10. Shrimps can crawl well, while prawns usually swim by rowing

Worksheet 6: Barnacles and Other Crustaceans; Spiders

1. False

2. True

3. False

4. True

5. True

6. False

7. True

8. False

9. True

10. True

Worksheet 7: Scorpions and Other Arachnids; Centipedes and Millipedes

1. Self-defense and sometimes to paralyze prey

2. Pseudoscorpions

3. Ticks and mites

4. They have one-piece bodies

5. Rocky Mountain fever and Lyme disease

6. Hundred legs

7. Around 40

8. Thousand legs

9. Around 750

10. Answers will vary.

SECTION 4: FISH

Worksheet 1: What Are Fish? Sharks and Rays

1. Lungfish

2. They use a "rod and line" made of their own body to catch other fish

3. Animals with backbones, and yes they are

4. Cold-blooded

5. They have a "backbone," or vertebral column, made of tough, gristly cartilage rather than true bone

6. A shark

7. No. They are covered in little pointed structures known as dermal denticles.

8. All chimaeras have a long spine at the front of the dorsal (back) fin. This is linked to a venom gland.

9. No.

10. The whale shark

Worksheet 2: Sturgeons and Gars; Eels and Herrings

1. True

2. False

3. True

4. True

5. True

6. False

7. True

8. False

9. True

10. False

Worksheet 3: Salmon, Pike, and Hatchetfish; Characins, Carp, and Catfish

1. In rivers and lakes

2. To prey on smaller fish

3. Silvery sea trout and the brown trout

4. The muskie or muskellunge of North America

5. They hide in weeds, lunge for it, and then capture it with their sharp teeth.

6. Nearly a third

7. No. They have a pair of toothed bones in their throats.

8. Characin

9. Long whisker-like barbels around their mouth

10. They have special muscles that can produce pulses of electricity in the surrounding water. These pulses help them to find their way in muddy, dark lakes and rivers. The electricity is so strong, it can stun prey fish.

Worksheet 4: Cod, Anglerfish, and Toadfish; Scorpionfish and Seahorses

1. Answers will vary, and can include: hake, haddock, and whiting

2. Fish that are not built for speed often have to find other methods of capturing their food instead of just chasing them down.

3. Because they look like toads and make croaks like a toad

4. Its body shape helps to camouflage it among the seaweeds

5. Long and slender and almost round as a balloon

6. Yes; other fish and shellfish

7. A long, tube-shaped snout

8. The strong spines on their backs

9. The male is the one that cares for the eggs while they hatch and grow.

10. To be dried and ground up as medicine, and as part of souvenirs

Worksheet 5: Flyingfish, Silversides, and Killifish Flatfish and Triggerfish

1. Flyingfish group

2. No — they glide above the surface of the water for a few second using their fins.

3. Its long snout is only made of its lower jaw.

4. North American Pacific coasts

5. Killfish group

6. The larvae and pupae of mosquitoes

7. No

8. The polar regions

9. Their mouth and a few teeth form a beak-like structure similar to a bird's beak.

10. Protection from predators

Worksheet 6: Perch, Groupers, and Drums; Tunas and Marlins

1. The perciforms

2. They ambush their prey by hiding in rocks and grabbing things in their large mouths.

3. Drum fish and croaker fish

4. 160

5. Yes — the brown marbled grouper is an example.

6. Plankton

7. Answers may include: tunas, mackerels, bonitos

8. Billfish

9. Yes, they can be.

10. It may help a swordfish when swimming, or to attack enemies, or as a weapon in hunting its prey.

Worksheet 7: Cichlids, Damsels, and Parrotfish; Blennies, Gobies, and Wrasses

1. Freshwater

2. In its throat and on its jaw

3. Clown fish

4. In shallow water

5. Alone

6. They are on stalks.

7. They often clean parasites off of larger fish.

8. Mudskippers

9. Everywhere but polar regions

10. The dwarf gobies in the Philippine Islands

SECTION 5: AMPHIBIANS AND REPTILES

Worksheet 1: What Are Amphibians? Newts and Salamanders

1. False
2. True
3. True
4. True
5. False
6. True
7. False
8. True
9. True
10. True

Worksheet 2: Frogs and Toads; What Are Reptiles?

1. Frogs and toads

2. 10 times

3. Plants

4. No. They differ by species.

5. Male

6. They are poisonous and deadly.

7. It uses its huge, web-like feet to glide.

8. Answers will vary and can include: by movement, skin crests, coloring, chemical messages, hisses, grunts, or calls

9. Yes

10. They have wide feet with partly webbed toes.

11. They cannot control their body temperature.

12. The slow movement or non-movement of reptiles due to cold conditions

13. They don't have to eat as much as warm-blooded animals.

Worksheet 3: Tortoises, Turtles, and Terrapins; Sea Turtles

1. By its hard shell

2. Chelonians

3. No

4. No. Some eat plants.

5. These are the two pieces of a turtle's shell.

6. It looks like a clump of bark and leaves in the water. When a victim passes by, the matamata simply opens its large mouth and the water and prey go in its mouth.

7. On the Galapagos group of islands

8. Sea or marine turtles

9. It is flatter and less domed

10. Answer will vary. Can include portions of the following: they have to break out of their eggs, make their way out of their sandpit nests, and then make a dash for the sea before they are eaten by birds or other predators. Even those that reach the sea try to reach deeper water, but can be eaten by fish.

Worksheet 4: Crocodiles and Alligators; How Crocodiles Breed

1. Crocodiles, alligators and caimain, and the gharial (gavial)

2. The fourth tooth is visible on the outside jaw of a crocodile but not an alligator.

3. At night

4. Smaller ones eat frogs, spiders, and insects; larger ones can eat birds, small mammals or fish.

5. A variety of sounds, smells, and movements

6. Yes

7. 60 to 100 days

8. In a nest she forms

9. In mounds made of dirt, plants, and dead leaves

10. The temperature

Worksheet 5: Iguanas, Agamids, and Chameleons; Geckos, Lacertids, and Teiid Lizards

1. Lizards

2. Tropical areas, but they are found everywhere but the far north and Antartica

3. On land and in trees

4. Most eat insects and other small creatures, but some eat plant material instead.

5. It lives mostly in water.

6. Agamid

7. Its feet are designed to grip trees, its tongue is sticky-tipped to trap prey, it can change color to camouflage itself, and has good eyesight.

8. No

9. Answers may vary — can include: the green and wall lizards, sandracer, and racerunner

10. No — it is a hynchocephalian

Worksheet 6: Skinks, Monitors, and Slow Worms; Pythons, Boas, and Thread Snakes

1. Skinks

2. Other lizards, snakes, small mammals, bird and lizard eggs

3. Amphisbaenids

4. Some do, and other do not

5. The Komodo dragon

6. Monitor lizards

7. Gila monster and beaded lizard

8. Antarctica

9. Underground

10. They coil themselves around their prey, crushing it until their prey suffocates.

11. No

12. Anaconda

Worksheet 7: Colubrid Snakes; Cobras, Vipers, and Rattlers

1. Colubrid snakes

2. Most are not but a few are

3. Yes

4. Back-fanged venomous snakes

5. Live young

6. Egg-eating snake of Africa

7. Cobras and vipers

8. They raise the front of their body off the ground, spread their ribs and the loose skin on the sides of the head, and they sway to look bigger.

9. Around 50

10. They have sensory pits that can sense the heat

that comes from warm-blooded prey.

SECTION 6: BIRDS

Worksheet 1: Flightless Birds; Seabirds

1. No
2. Puffins, razorbills, and guillemots
3. Warm-blooded
4. The dodo bird
5. Oil
6. About one pound
7. Ratites
8. Albatrosses, shearwaters, and petrels
9. Around 90
10. Auks
11. Answers will vary, may include: ostriches, emus, rheas, dodos, kakapo parrot, kiwis, cassowaries, penguins
12. Answers will vary, may include: terns, auks, skuas, frigate birds, gulls, terns, skimmers, petrels, guillemots, cormorants, shearwaters

Worksheet 2: Shorebirds and Waterbirds; Herons, Ducks, and Geese

1. The edge of the water
2. On the water, in reeds and vegetation, on the shore
3. Long legs and wide-spread toes that help them walk through mud and water, beaks that are long and thin to poke about and dig in the sand for things to eat
4. Almost twice as much
5. Grebes
6. To pry apart shellfish like mussels
7. Swans
8. Ibises
9. The male bittern
10. Yes

Worksheet 3: Birds of Prey; Gamebirds and Rails

1. Sharp, hooked beak and curved claws
2. They can stay aloft for house, keen eyesight, and dive quickly
3. Rats, rabbits, snakes, lizards, and other birds
4. Eagle; power
5. Hawks, sparrowhawks, goshawks, falcons, kestrels, hobbies
6. Accipiters and buteos
7. Many large birds that live mainly on the ground and are hunted for meat
8. Answers will vary; can include pheasants, grouse, partridges, capercaillies, quails, wild turkeys, guinea fowl, curassows, guans; they are often camouflaged or colored to blend in
9. 165
10. Rails

Worksheet 4: Pigeons, Doves, and Parrots; Cuckoos and Turacos

1. They eat mostly seeds and fruits; they are bold and curious, placid and easy-going and have long associations with people.
2. The parrot group
3. Doves and pigeons
4. Untidy stick nests
5. Parrots
6. In tree trunks in Central and South America
7. In the nests of other birds
8. From its call
9. Fruit
10. Roadrunner

Worksheet 5: Owls and Nightjars; Swifts and Woodpeckers

1. Owls, nightjars, and frogmouths
2. They can see in almost pitch darkness.
3. Four times
4. Moths and night-flying insects
5. Owl
6. Up to 80 times
7. They are so well-designed for flight they rarely

are not in the air.

8. Swifts

9. Over 300 species and they eat flower nectar

10. The same type of feet with two toes pointing forward and two toes pointing backward

Worksheet 6: Crows, Shrikes, and Bowerbirds; Sparrows, Finches, and Weavers

1. True
2. False
3. True
4. True
5. True
6. False
7. False
8. True
9. False
10. True

Worksheet 7: Warblers, Thrushes, and Flycatchers; Larks, Swallows, and Treecreepers

1. A call is a short message relaying information; songs are usually only done by males during the mating season
2. Their drab-colored plumage
3. In warmer climates to the south
4. Thrushes
5. The marsh warbler
6. No — the smallest
7. Under the eaves of houses
8. To easily pick up the insects that they eat
9. Dippers
10. Cup-like

SECTION 7: MAMMALS

Worksheet 1: Egg-Laying Mammals; Marsupial Mammals

1. False
2. False

3. True
4. True
5. True
6. True
7. False
8. True
9. True
10. False

Worksheet 2: Insect Eaters; Bats

1. Insectivores
2. Nocturnal
3. They lose more body heat through the surface o the body and eat lots of food to keep their energ and temperature up.
4. About 5,000
5. Every few hours
6. Answers may vary, can include: shrews, hedgehogs, moles, tenrecs, solenodons, moonrats, and desmans
7. About one-fourth
8. Yes
9. Fruit bats and flying foxes
10. A sound-radar system called echolocation
11. Their huge ears which can turn and swivel
12. Common bats or vesper bats
13. Yes, but they are not dangerous to people as the feed on animals

Worksheet 3: Anteaters, Sloths, and Armadillos; Rabbits, Hares, and Pikas

1. They live at a slow pace with long sharp claws, very small or no teeth at all
2. Ants, termites, and small grubs
3. Four
4. They have a covering of tough bone or horn-like armor
5. They are so very slow
6. Pangolins

. Their back legs are long and muscular.

. There really isn't one — except for size; smaller species are called rabbits

. They have strong teeth.

0. Young hares

Worksheet 4: Hyraxes and the Aardvark; Mice, Rats, and Cavies

. Hyraces

. Rock hydraxes

. Aardvark

. Between cracks in rocks, in caves, under tree roots or in the hollow of trees

. Narrow head, with long ears, and a long flexible snout with a pig-like nose

. Rodents or gnawing mammals

. They can cause damage and spread disease.

. Stretchy cheek pouches

. The deserts of North Africa, the Middle East, and Central Asia

0. Common dormice

Worksheet 5: Large Rodents; Squirrels and Chipmunks

. True

. False

. True

. False

. True

. True

. False

. True

. True

0. True

Worksheet 6: Deer, Camels, and Pigs; Antelopes, Wild Cattle, and Sheep

. Ungulates

. Camels

3. Bactrian (two-humped) and dromedary (one-humped)

4. Deer

5. Reindeer or caribou

6. For safety from predators

7. Antelopes and cattle

8. Tibet

9. A thick, warm coat

10. Their speed and agility as well as keen senses

Worksheet 7: Horses, Zebras, and Rhinos; Elephants

1. Answers can vary — may include wild horses, zebra, rhinos, and tapirs

2. Plants

3. Southeast Asia

4. Striped horses

5. Tightly packed hairs

6. Elephants

7. In groups

8. Around 18

9. No — they suck it up with the trunk and then squirt the water into their mouths

10. Its incisor teeth

Worksheet 8: Cats; Dogs, Foxes, and Hyenas

1. Similar in overall body shape and features, as well as in their hunting methods

2. Meat-eater

3. Mammals with long, sharp claws to catch prey, and long, sharp teeth to eat it

4. Answer will vary — may include lions, cats, lynx, wolves, wild dogs, hyenas, bears, raccoons, panda, civets, genets, weasels, mongooses, seals, and sea lions

5. Other carnivores may scavenge or eat fruit occasionally, but cats are active hunters of living prey.

6. The cheetah

7. Lions, pride

8. The mountains of Asia

9. A once-tame cat that has returned to living in the wild

10. Dogs have longer muzzles; they cannot pull their claws back into their toes, and most live in pairs or groups called packs (except for the fox that lives alone)

11. The gray wolf

12. It changes to white in the winter and helps to camouflage it from predators, but also helps it be able to capture prey

13. Dogs can run for long distances and their sense of smells enables them to track prey until it tires and can be caught.

14. Canis familiaris

15. Fox

Worksheet 9: Bears, Raccoons, and Pandas; Weasels, Mongooses, and Civets

1. The giant panda

2. Asia

3. The white fur helps camouflage it when hunting seals, and a thick layer of fat keeps out the cold; they can also run fast and have a good sense of smell.

4. Fish, frogs, small birds, mammals, eggs, fruits, nuts, seeds . . . just about anything

5. Its prehensile tail works like a fifth limb.

6. Answers will vary and can include any number of the following: weasels, stoats, skunks, badgers, mink, otters, mustelids, mongooses, civets, genets, linsangs.

7. Skunk

8. Wolverines

9. They eat mice, rats, small insects, and even poisonous snakes.

10. Answers can vary and may include some of the following: they have webbed feet to paddle with and a tail to help them move faster in the water, their fur is waterproofed, and the thick underfur even traps bubbles of air so the otter can stay warm

Worksheet 10: Seals, Sea Lions, and Sea Cows

1. Near land, on coasts, or near the ice

2. Water plants; manatee and dugongs

3. The leopard seal

4. Fish and squid

5. Pups

6. It doesn't eat crabs.

7. True or earless and eared seals

8. They cannot move quickly on land.

9. Rookeries

10. Sea cows

Worksheet 11: Great Whales

1. Only on accident when they are beached

2. They have comb-like baleen in their mouths for filter-feeding.

3. Being hunted in large numbers

4. Harpoon

5. Pod

6. Humpback whale

7. Calves

8. Migration

9. Around 300 feet

10. Blue and humpback whales take a massive mouthful of sea water; they press the huge tongue upward inside the mouth to force the water out through the baleen plates. Food items such as krill and small fish get trapped in the combs or bristles of the baleen, which are tough and springy, like plastic. The whale then licks the food off the baleen plates and swallows it.

Worksheet 12: Dolphins and Porpoises

1. Answers may vary — can include porpoises and dolphins, sperm whales, beaked whales, narwhal and beluga, killer whales, etc.

2. Sperm whale

3. They will sometimes work together in herding a school of fish to shallow water to eat them.

- Giant squid
- A dolphin
- Use a great variety of squeals, buzzes, clicks, and grunts
- A sword-like tusk
- Rivers
- A dolphin has a protruding beak-like snout and a curving sickle-shaped back fin. Porpoises are similar, but lack the dolphin's beak and have a snub nose instead. They are also generally smaller than dolphins.
10. The narwhal and beluga

Worksheet 13: Lemurs and Bushbabies; African Monkeys

- Primates
- Well-developed hands with thumbs that can grasp, and large, forward facing eyes.
- Madagascar
- Early-monkeys
- Smaller than monkeys with simple teeth and few points (cusps).
- Mangabeys
- Colobus monkeys
- In the trees
- Habitat loss
10. Their very loud communication

Worksheet 14: Asian Monkeys; American Monkeys

- True
- True
- False
- False
- True
- True
- True
- False
- True
10. False

Worksheet 15: Gibbons; Orangs and Gorillas

1. Yes
2. 9
3. The female
4. Kloss gibbon
5. 40 years
6. Orang, gorilla, chimp, and pygmy chimp
7. Man of the woods
8. Answers will vary. Should include details like: gorillas are very heavily built, with a broad chest, long arms, and a large, tall-domed head with massive jaws and teeth. The nose is black, leathery, and flattened. The fur and skin is mostly black, except in older adult males, which develop a silvery-white patch on the back and flanks, giving them the name of silverbacks.
9. Gorillas
10. Orangutan

Worksheet 16: Chimpanzees

1. True
2. True
3. False
4. True
5. False
6. False
7. True
8. True
9. False
10. True

Introduction/ Section 1: Simple Animals Quiz #1

1. Monera, Protista, Plants, Fungi, and Animals
2. Animals eat other living things; plants get their energy from sunlight.
3. Vertebrates and invertebrates
4. If it has very few basic body parts
5. Sponges
6. They thrive inside other living things, known as hosts.
7. Through mosquito bites
8. It uses sunlight energy.
9. They produce food for themselves rather than having to consume it.
10. They produce egg-like structures and are often blown to different places by the wind, starting new colonies.
11. Wheel animals
12. They filter water, helping keep it clean and clear.
13. They drift or swim with the currents.
14. The ability of an animal to create their own light
15. Fish, squid, and sea turtles
16. Sense organs that can detect light or dark, as well as which way is up or down
17. Sea anemones
18. Trap fish and other small creatures in their tentacles
19. A helpful partnership between two different kinds of animals that is beneficial to both of them
20. Stony, soft, gorgonian

Section 2: Worms, Snails, & Starfish Quiz #2

1. Most worms live only in water or under the ground in damp mud, sand, or soil.
2. Worms absorb oxygen through their thin, moist skin.
3. Yes, they are like flattened types of worms. Some leeches eat smaller worms or insect grubs, others eat blood from fish and even people.
4. Flatworms; no ring-like sections or segments
5. Protects it from the body fluids and digestive juices from the animal it lives within
6. With hooks and suckers
7. Answers can include: ribbon worms, spoon worms, arrow worms, acorn worms, horsehair worms, peanut worms, beard worms, horseshoe worms, priapulids, tongue worms, spiny-headed worms
8. Ribbon worms
9. Mollusks
10. Answers can include: to catch prey, to cling to rocks, to slide around, to dig around in sand or mud.
11. It is like a trapdoor that protects the snail in its shell.
12. Sea snails are often very colorful, while land snails are mostly gray and brown.
13. On land they have lungs; those in the sea have gills.
14. Answers will vary but can include: tower shells, top shells, cowries, conches
15. With two valves
16. A bit of grit or a parasite may enter the shell. If the mollusk cannot get rid of it, it is gradually covered in nacre, the substance that forms the smooth, white inner lining of a shell. This builds up in thin layers until the grit is completely wrapped in a ball of shiny white nacre forming a pearl.
17. They are able to change both the color and patterns on their skin quickly, easily, and often.
18. Hedgehog skin
19. In the sea
20. Mollusks, bivalves, crabs, worms, and other echinoderms

Section 3: Insects & Other Arthropods #3

1. Insects
2. Hard outer body casing or exoskeleton, wings for flying, and six legs with flexible joints for running

. Eight

. Arthropods

. Answer will vary; may include: ants, beetles, bugs, cockroaches, flies, fleas, bees, termites, locusts, and other insects, spiders and scorpions, the multi-legged centipedes and millipedes, and crustaceans such as crabs, lobsters, and shrimps

. Four

. Termites

. One pair of proper wings, with a small second pair in the back that help them to balance

. No

0. False

1. False

2. True

3. Hosts

4. Crustaceans

5. False

6. False

7. True

8. Self-defense and sometimes to paralyze prey

9. They have one-piece bodies

0. Hundred legs

ection 4: Fish Quiz #4

. Lungfish

. They use a "rod and line" made of their own body to catch other fish

. Animals with backbones, and yes they are

. Cold-blooded

. True

. False

. To prey on smaller fish

. Silvery sea trout and the brown trout

. Nearly a third

0. Long whisker-like barbels around their mouth

1. A long, tube-shaped snout

2. The male is the one that cares for the eggs while they hatch and grow.

13. Flyingfish group

14. Killfish group

15. The perciforms

16. Plankton

17. Billfish

18. Freshwater

19. Clown fish

20. They often clean parasites off of larger fish.

Section 5: Amphibians and Reptiles Quiz #5

1. False

2. True

3. True

4. Frogs and toads

5. Male

6. Answers will vary and can include: by movement, skin crests, coloring, chemical messages, hisses, grunts, or calls

7. They cannot control their body temperature.

8. The slow movement or non-movement of reptiles due to cold conditions

9. By its hard shell

10. These are the two pieces of a turtle's shell.

11. Answer will vary. Can include portions of the following: they have to break out of their eggs, make their way out of their sandpit nests, and then make a dash for the sea before they are eaten by birds or other predators. Even those that reach the sea try to reach deeper water, but can be eaten by fish.

12. The fourth tooth is visible on the outside jaw of a crocodile but not an alligator.

13. Smaller ones eat frogs, spiders, and insects; larger ones can eat birds, small mammals or fish.

14. The temperature

15. Lizards

16. Its feet are designed to grip trees, its tongue is sticky-tipped to trap prey, it can change color to camouflage itself, and has good eyesight.

17. No — it is a hynchocephalian

18. Skinks

19. The Komodo dragon

20. Antarctica

21. No

22. Colubrid snakes

23. Back-fanged venomous snakes

24. Live young

25. Cobras and vipers

Section 6: Birds Quiz # 6

1. No

2. Warm-blooded

3. The dodo bird

4. Long legs and wide-spread toes that help them walk through mud and water, beaks that are long and thin to poke about and dig in the sand for things to eat

5. Swans

6. Yes

7. They can stay aloft for hours, keen eyesight, and dive quickly

8. Eagle; power

9. Many large birds that live mainly on the ground and are hunted for meat

10. The parrot group

11. Parrots

12. In the nests of other birds

13. They can see in almost pitch darkness.

14. Four times

15. Up to 80 times

16. Swifts

17. False

18. False

19. True

20. Under the eaves of houses

Section 7: Mammals: Part One Quiz #7

1. False

2. True

3. True

4. True

5. Nocturnal

6. They lose more body heat through the surface of the body and eat lots of food to keep their energy and temperature up.

7. Yes

8. They have a covering of tough bone or horn-like armor

9. They are so very slow

10. There really isn't one — except for size; smaller species are called rabbits

11. Hyraces

12. They can cause damage and spread disease.

13. Stretchy cheek pouches

14. Common dormice

15. True

16. True

17. False

18. True

19. Ungulates

20. Bactrian (two-humped) and dromedary (one-humped)

21. For safety from predators

22. A thick, warm coat

23. Striped horses

24. In groups

25. No — they suck it up with the trunk and then squirt the water into their mouths

Section 7: Mammals: Part Two Quiz #8

1. Meat-eater

2. Other carnivores may scavenge or eat fruit occasionally, but cats are active hunters of living prey.

3. A once-tame cat that has returned to living in the wild

4. Canis familiaris

5. The white fur helps camouflage it when hunting seals, and a thick layer of fat keeps out the cold;

they can also run fast and have a good sense of smell.

. Answers will vary and can include any number of the following: weasels, stoats, skunks, badgers, mink, otters, mustelids, mongooses, civets, genets, linsangs.

. They eat mice, rats, small insects, and even poisonous snakes.

. Water plants; manatee and dugongs

. Pups

0. True or earless and eared seals

1. They cannot move quickly on land.

2. Being hunted in large numbers

3. Humpback whale

4. Calves

5. Blue and humpback whales take a massive mouthful of sea water; they press the huge tongue upward inside the mouth to force the water out through the baleen plates. Food items such as krill and small fish get trapped in the combs or bristles of the baleen, which are tough and springy, like plastic. The whale then licks the food off the baleen plates and swallows it.

16. Sperm whale

17. Use a great variety of squeals, buzzes, clicks, and grunts

18. A dolphin has a protruding beak-like snout and a curving sickle-shaped back fin. Porpoises are similar, but lack the dolphin's beak and have a snub nose instead. They are also generally smaller than dolphins.

19. Primates

20. True

21. True

22. Yes

23. Man of the woods

24. True

25. True

The World of Animals 🔑 Test Answer Keys

. Monera, Protista, Plants, Fungi, and Animals

. Animals eat other living things; plants get their energy from sunlight.

. Vertebrates and invertebrates

. Sponges

. They produce food for themselves rather than having to consume it.

. They filter water, helping keep it clean and clear.

. The ability of animals to create their own light

. A helpful partnership between two different kinds of animals that is beneficial to both of them

. Most worms live only in water or under the ground in damp mud, sand, or soil.

0. Worms absorb oxygen through their thin, moist skin.

1. Answers can include: ribbon worms, spoon worms, arrow worms, acorn worms, horsehair worms, peanut worms, beard worms, horseshoe worms, priapulids, tongue worms, spiny-headed worms

12. On land they have lungs; those in the sea have gills.

13. With two valves

14. Hedgehog skin

15. Hard outer body casing or exoskeleton, wings for flying, and six legs with flexible joints for running

16. No

17. Hosts

18. Crustaceans

19. They have one-piece bodies

20. Animals with backbones, and yes they are

21. Cold-blooded

22. A long, tube-shaped snout

23. The perciforms

24. Freshwater

25. Frogs and toads

26. They cannot control their body temperature.

27. The fourth tooth is visible on the outside jaw of a crocodile but not an alligator.

28. Smaller ones eat frogs, spiders, and insects; larger ones can eat birds, small mammals or fish.

29. Its feet are designed to grip trees, its tongue is sticky-tipped to trap prey, it can change color to camouflage itself, and has good eyesight.

30. Colubrid snakes

31. Live young

32. No

33. Warm-blooded

34. They can stay aloft for hours, keen eyesight, and dive quickly

35. Parrots

36. They can see in almost pitch darkness.

37. False

38. True

39. True

40. Nocturnal

41. They lose more body heat through the surface of the body and eat lots of food to keep their energy and temperature up.

42. Yes

43. Ungulates

44. Meat-eater

45. A once-tame cat that has returned to living in the wild

46. The white fur helps camouflage it when hunting seals, and a thick layer of fat keeps out the cold; they can also run fast and have a good sense of smell.

47. True or earless and eared seals

48. Humpback whale

49. Use a great variety of squeals, buzzes, clicks, and grunts

50. True

Now turn your favorite **Master Books** into curriculum! Each Parent Lesson Plan (PLP) includes:

- An easy-to-follow, one-year educational calendar
- Helpful worksheets, quizzes, tests, and answer keys
- Additional teaching helps and insights
- Complete with all you need to quickly and easily begin your education program today!

ELEMENTARY ZOOLOGY

1 year
4th – 6th

Package Includes: *World of Animals, Dinosaur Activity Book, The Complete Aquarium Adventure, The Complete Zoo Adventure, Parent Lesson Plan*

5 Book Package
978-0-89051-747-5 $84.99

SCIENCE STARTERS: ELEMENTARY PHYSICAL & EARTH SCIENCE

1 year
3rd – 8th grade

6 Book Package Includes: *Forces & Motion – Student, Student Journal, and Teacher; The Earth – Student, Teacher & Student Journal; Parent Lesson Plan*

6 Book Package
978-0-89051-748-2 $51.99

SCIENCE STARTERS: ELEMENTARY CHEMISTRY & PHYSICS

1 year
3rd – 8th grade

Package Includes: *Matter – Student, Student Journal, and Teacher; Energy – Student, Teacher, & Student Journal; Parent Lesson Plan*

7 Book Package
978-0-89051-749-9 $54.99

INTRO TO METEOROLOGY & ASTRONOMY

1 year
7th – 9th grade
½ Credit

Package Includes: *The Weather Book; The Astronomy Book; Parent Lesson Plan*

3 Book Package
978-0-89051-753-6 $44.99

INTRO TO OCEANOGRAPHY & ECOLOGY

1 year
7th – 9th grade
½ Credit

Package Includes: *The Ocean Book; The Ecology Book; Parent Lesson Plan*

3 Book Package
978-0-89051-754-3 $45.99

INTRO TO SPELEOLOGY & PALEONTOLOGY

1 year
7th – 9th grade
½ Credit

Package Includes: *The Cave Book; The Fossil Book; Parent Lesson Plan*

3 Book Package
978-0-89051-752-9 $44.99

CONCEPTS OF MEDICINE & BIOLOGY

1 year
7th – 9th grade
½ Credit

Package Includes: *Exploring the History of Medicine; Exploring the World of Biology; Parent Lesson Plan*

3 Book Package
978-0-89051-756-7 $40.99

CONCEPTS OF MATHEMATICS & PHYSICS

1 year
7th – 9th grade
½ Credit

Package Includes: *Exploring the World of Mathematics; Exploring the World of Physics; Parent Lesson Plan*

3 Book Package
978-0-89051-757-4 $40.99

CONCEPTS OF EARTH SCIENCE & CHEMISTRY

1 year
7th – 9th grade
½ Credit

Package Includes: *Exploring Planet Earth; Exploring the World of Chemistry; Parent Lesson Plan*

3 Book Package
978-0-89051-755-0 $40.99

THE SCIENCE OF LIFE: BIOLOGY

1 year
8th – 9th grade
½ Credit

Package Includes: *Building Blocks in Science; Building Blocks in Life Science; Parent Lesson Plan*

3 Book Package
978-0-89051-758-1 $44.99

BASIC PRE-MED

1 year
8th – 9th grade
½ Credit

Package Includes: *The Genesis of Germs; The Building Blocks in Life Science; Parent Lesson Plan*

3 Book Package
978-0-89051-759-8 $43.99

INTRO TO ASTRONOMY

1 year
7th – 9th grade
½ Credit

Package Includes: *The Stargazer's Guide to the Night Sky; Parent Lesson Plan*

2 Book Package
978-0-89051-760-4 $47.99

INTRO TO ARCHAEOLOGY & GEOLOGY

1 year
7th – 9th
½ Credit

Package Includes: *The Archaeology Book; The Geology Book; Parent Lesson Plan*

3 Book Package
978-0-89051-751-2 $45.99

SURVEY OF SCIENCE HISTORY & CONCEPTS

1 year
10th – 12th grade
1 Credit

Package Includes: *The World of Mathematics; The World of Physics; The World of Biology; The World of Chemistry; Parent Lesson Plan*

5 Book Package
978-0-89051-764-2 $72.99

SURVEY OF SCIENCE SPECIALTIES

1 year
10th – 12th grade
1 Credit

Package Includes: *The Cave Book; The Fossil Book; The Geology Book; The Archaeology Book; Parent Lesson Plan*

5 Book Package
978-0-89051-765-9 $81.99

SURVEY OF ASTRONOMY

1 year
10th – 12th grade
1 Credit

Package Includes: *The Stargazers Guide to the Night Sky; Our Created Moon; Taking Back Astronomy; Our Created Moon DVD; Created Cosmos DVD; Parent Lesson Plan*

4 Book, 2 DVD Package
978-0-89051-766-6 $112.99

GEOLOGY & BIBLICAL HISTORY

1 year
8th – 9th
1 Credit

Package Includes: *Explore the Grand Canyon; Explore Yellowstone; Explore Yosemite & Zion National Parks; Your Guide to the Grand Canyon; Your Guide to Yellowstone; Your Guide to Zion & Bryce Canyon National Parks; Parent Lesson Plan*

4 Book, 3 DVD Package
978-0-89051-750-5 $112.99

PALEONTOLOGY: LIVING FOSSILS

1 year
10th – 12th grade
½ Credit

Package Includes: *Living Fossils, Living Fossils Teacher Guide, Living Fossils DVD; Parent Lesson Plan*

3 Book, 1 DVD Package
978-0-89051-763-5 $66.99

LIFE SCIENCE ORIGINS & SCIENTIFIC THEORY

1 year
10th – 12th grade
1 Credit

Package Includes: *Evolution: the Grand Experiment, Teacher Guide, DVD; Living Fossils, Teacher Guide, DVD; Parent Lesson Plan*

5 Book, 2 DVD Package
978-0-89051-761-1 $139.99

NATURAL SCIENCE THE STORY OF ORIGINS

1 year
10th – 12th grade
½ Credit

Package Includes: *Evolution: the Grand Experiment; Evolution: the Grand Experiment Teacher's Guide, Evolution: the Grand Experiment DVD; Parent Lesson Plan*

3 Book, 1 DVD Package
978-0-89051-762-8 $66.99

ADVANCED PRE-MED STUDIES

1 year
10th – 12th grade
1 Credit

Package Includes: *Building Blocks in Life Science; The Genesis of Germs; Body by Design; Exploring the History of Medicine; Parent Lesson Plan*

5 Book Package
978-0-89051-767-3 $76.99

BIBLICAL ARCHAEOLOGY

1 year
10th – 12th grade
1 Credit

Package Includes: *Unwrapping the Pharaohs; Unveiling the Kings of Israel; The Archaeology Book; Parent Lesson Plan.*

4 Book Package
978-0-89051-768-0 $99.99

CHRISTIAN HERITAGE

1 year
10th – 12th grade
1 Credit

Package Includes: *For You They Signed; Lesson Parent Plan*

2 Book Package
978-0-89051-769-7 $50.99